Giulia **Parola**

Environmental Democracy at the Global Level:
Rights and Duties for a New Citizenship

Versita Discipline:
Language, Literature

Managing Editor:
Margherita Poto

Language Editor:
Laura Isakoff
Sara Suliman

Published by Versita, Versita Ltd, 78 York Street, London W1H 1DP, Great Britain.

ISBN: (hardcover): 978-83-7656-013-7

ISBN (paperback): 978-83-7656-012-0

ISBN (for electronic copy): 978-83-7656-014-4

Managing Editor: Margherita Poto

Language Editor: Laura Isakoff
                 Sara Suliman

Cover illustration: ©iStockphoto.com/blackred

www.versita.com

To "La Ca di Banda", where I founded my roots,
to my unique and special father Sergio.

Contents

Abbreviations

BoE	- Bank of England Archives, London, Great Britain
C	- Case
CFCA	- Community Fisheries Control Agency
CFI	- Court First Instance
COM	- Communication
CONF	- Conference
CSOs	- Civil Society Organisations
Doc.	- Document
e.g.	- Exempli Gratia
EC	- European Community
ed.	- Editor
eds.	- Editors
EIA	- environmental Impact Assessment
ETS	- European Treaty Series
EU	- European Union
f.	- Following
FAO	- Food and Agricultural Organisation
G.A.	- General Assemble
ICJ	- International Court of Justice
IPCC	- International Panel on Climate Change
IPPC	- Integrated Pollution Prevention and Control
ISSN	- International Standard Serial Number
IUCN	- International Union for Conservation of Nature
MS	- Members States
NGO	- Non Governmental Organisation
OECD	- Organisation for Economic Co-operation and Development
OJ	- Official Journal
OSCE	- Organisation for Security and Cooperation in Europe
p.	- Page
Para.	- Paragraph

pp.	- Pages
REACH	- Registration, Evaluation, Authorisation and Restriction of Chemicals
RMA	- Resource Management Act
S.I.	- Statutory Instruments
TFAJ	- Task Force on Access to Justice
TFEU	- Treaty on the Functioning of the European Union
UK	- United Kingdom
UN	- United Nations
UN/ECE	- United Nations Economic Commission for Europe
UNCED	- United Nations Conference on Environment and Development
UNECE	- United Nations Economic Commission for Europe
UNEP	- United Nations Environment Programme
UNESCO	- United Nations Educational, Scientific and Cultural Organisation
UNGA	- United Nations General Assembly
UNTS	- United Nations Treaties Series
UNU-IAS	- United Nations University-Institute of Advanced Studies (UNU-IAS)
US	- United States
Vol.	- Volume
WFD	- Water Framework Directive
WWF	- World Wildlife Fund
WWF-EPO	- WWF European Union environmental Policy

Introduction

"Il n'y a pas seulement pour l'humanité la menace de disparaître sur une planète morte. Il faut aussi que chaque homme, pour vivre humainement, ait l'air nécessaire, une surface viable, une éducation, un certain sens de son utilité. Il lui faut au moins une miette de dignité et quelques simples bonheurs"
(Yourcenar, 1980, p. 5).

1 The Challenge of the *Coeur* of the Book

There is only one Earth, and it is endangered. It has already survived ages of natural destruction and reconstruction; however, for the first time, the present dangers Earth is facing do not derive from a natural revolution, but from a whole variety of human activities, which have pushed the Planet to its limits (Speth,2008).

Furthermore, our age is characterised by a global ecological crisis; indeed, since the largest problem is the environmental one, every other issue is in one way or another, linked to this crisis. For instance, wars are slowly but surely becoming environmental conflicts for the control of food and water, as well as being a cause of increased poverty. The negation of fundamental human rights stems more and more often from environmental issues[1]. Finally, the economic crisis was provoked by irresponsible management of natural resources.[2]

---

[1] For instance the problem of the environmental Refugees. This topic will be briefly analysed in Chapter I

[2] For instance the petroleum. Commoner, 1992, p. 15.

Since immemorial time, humans have depended on and struggled against nature: they were both the causes and victims of environmental degradation (Bjerler, 2009). Indeed, no other living creature is so desperately dependent on nature, and thus so overtly vulnerable to any environmental change that occurs. No other species possesses such extensive capability to pollute and destroy the environment.

Initially, Earth was characterised by dynamic equilibrium and a number of biogeochemical cycles, with humans as an integral part of this system. But times have changed; humans have developed science and technology, and created an alternative world, which has been called "technosphere" (Commoner, 1992, p.3).

Humans created a new world within the biosphere, with its own processes and events, which now does not fit into the above-mentioned system of cycles and equilibrium. The *technosphere* has been placing too much pressure on the functions of Earth (Vonkeman, 1997, p. 319), leaving a deep footprint on its face and seriously endangered the fragile biological balance. The human attack on the biosphere has started an "ecological counterattack" by nature through climate change and natural disasters, shortages, rising sea levels, severe droughts and a decreased ability to provide for basic human needs (Commoner, 1992, p. 7). The "two worlds are at war"; an ecological war (Commoner, 1992, p. 7).

Legal instruments and the legal approach have also contributed to unleash this conflict. The first cause is the failure of natural resources law and environmental law, which were never fully aligned with ecological reality. The laws have addressed environmental problems by compartmentalising resources into separate categories according to each legal treatment and without taking into account the biological principles. These biological principles recognise the full ecology of nature and the interrelationships among all elements (Noss, 1994, p. 893). In fact, in the biosphere "groundwater is connected to surface water, migratory birds are dependent on water areas and forests. The forests areas are vital to the carbon cycle, and so is the entire workings of nature, operate together as a full ecology system" (M. C. Wood, 2009, p. 43; M. C. Wood, 2009, p. 91).

The second factor is the government's failure to protect natural resources on behalf of its citizens. The ecological problem has pointed out the weakness of this political system. The flop of "Copenhagen" in December 2009 has confirmed the truth of the above statement. Consequently, the Nation State, in particular national democracy, does not effectively address the ecological crisis.

Does this mean that the choice of democracy is wrong? This political model has to find a new, yet a traditional, ground. The delegation of power is not a solution anymore; the ecological issues are so complex and embedded so deeply in human life, that it has become necessary to rediscover the idea that democracy is government by the people.

A healthy environment is for the public good and a goal that must be protected against deterioration - through appropriate regulations which reflect the views of citizens directly involved in public decision-making to protect their stake, in others words the Earth's stake (Mathiesen, 2003, p. 38).

Thus, if the environment is polluted, suffering from bad management and a negligent attitude by former and present governments and generations, a theoretical possible solution to achieve ecological aims could be found through a radical change of political and legal structures of power, as well as, in a widespread alteration in the behaviour of individuals.

Therefore a variety of questions arise: How can the ecological war be stopped? How can the ecological crisis be resolved?

Alternatively, from a legal point of view, the question could be how political and legal structures could contribute to avoiding environmental damage and threats of an ecological crisis. How could legal structures begin to reform and restructure actual political institutions so that they are more in line with environmental considerations?

How can states and their citizens act and organise themselves to answer to the current ecological crisis?

While commentators in the economic arena increasingly suggest alternative economic models such as "natural capitalism" (Hawken, Lovins, & Lovins, 2000), the legal theory, unfortunately, is at best a clumsy institution to effectuate massive change, as it lacks innovative thinking.

This book suggests the construction of an environmental democracy, which would address the aforementioned questions. This proposal is a traditional response, since it underpins the power of a democratic government, on the other hand it is "revolutionary". This proposal provides, first, a necessary normative shift from a human-centred to an Earth-centred approach, where every governmental decision should consider and value every possible impact on the environment using a more eco-centric approach, where the short-term considerations of human welfare must be balanced with the long-term interests of Earth. Second, such process of assessment should include individuals and their functions to exercise control over acts of government, thereby participating and contributing to decision-making in environmental matters.

This form of democracy creatively and constructively involves all the voices of the community. Such effort means revitalising democracy to meet the ecological challenges and change the role of human "from conqueror of the land-community to a plain member and citizen of it" (Leopold, 1949, p. 204).

Further, environmental democracy is more than the balance of the good of humanity and of Earth. It is more than a path for better management of the behaviour of technology and civilisation; it is a way to build a new civilisation.

In order to move in this direction, it is important to develop strategies for modifying human behaviour towards environmentally benign practices and away from environmentally damaging ones. Law is an important tool since it creates legal frameworks for environmental rights and ecological duties, which lead each individual as a citizen of social and ecological communities to become aware of the incredibly powerful role they can have in this crisis.3

These rights enable individuals to make choices and exercise their power in their everyday lives in addressing environmental matters. These rights even at the lowest level include environmental goods such as clean air and water, but also extend to procedural rights to be included in decision-making about the environment (WHAT, 2000, p. 7). This stakeholder approach acknowledges not only ecological duties vis-à-vis present and future generations, but also the ecological wellbeing of the Earth.

The mentioned rights and duties can only emerge through a legal process, which according to this book can be achieved through a construction of environmental democracy.

Hence, it is necessary to mobilise all mentioned rights and duties to make the transition from, in Bosselmann's words "Homo economicus to Homo ecologicus", to eventually justify the name Homo sapiens" (Bosselmann, 2009, p. 330).

What may be needed now is a clear vision of what the new form of democracy might look like.

2 A Theoretical Solution

Once the causes of the ecological crisis have been clarified, it is necessary to return to a legal point of view and try to link different points expressed above, some of which are not purely legal but are nevertheless the basis of such legal issues.

The idea of an environmental democracy comes from the attempt to seek a theoretical legal solution without twisting the political system and finding a different way to use the democratic concepts and tools.

In order to achieve this objective, Chapter I **"Environmental Democracy: A Theoretical Construction"** presents the conceptual building blocks of this book's approach, suggesting the possible transformation of the actual political and legal structures into an "environmental democracy".

Before speaking about the elements – form, space and actors – which compose the environmental democracy, it is necessary to analyse in Section I of this

---

3 Some scholars have suggested techniques for modifying human behaviour can be thought of as falling into two types: incentives and disincentives. Wilkinson, 2002, p. 10. See also Pathak, 1992, p. 205-206.

Chapter, titled **"Environmental Democracy"**, what the notions of "democracy" and "Environment" in this book's prospective encompass.

This section deals with two main issues: the first explores different theoretical forms of democracy, in particular, the participative and deliberative theories, which are both fundamental for the purpose of this book and the meaning of the term "environment". The second is to unify the two analysed notions into one concept, called "environmental democracy", and to contribute to a construction of a form and spatial dimension of this new type of democracy.

This section will argue that an ideal form of environmental democracy should include elements of deliberative and participatory democracy, as well as their processes and mechanisms where the citizens have a real possibility to participate. There seems to be a generally accepted view that the public should be involved directly in environmental decision-making. Hence, the emphasis on public democratic participation, participatory and deliberative, is the most significant theoretical solution proposed to answer to "certain disillusionment with the authority of the state to regulate for environmental protection", and is being more and more mirrored in International, European and domestic environmental law (Lee, & Abbot, 2003, p. 80).

From a spatial perspective, environmental democracy should be set up at the global and local levels to manage global and local ecological problems. Firstly, this takes place through international environmental law, and secondly, through regional and national regulation (Newigl, & Fritsch, 2009, p. 197).

The second point, which will be studied in the Section II of Chapter I, called **"The Actors of Environmental Democracy: The Environmental and Ecological Citizen"**, is related to the actors of environmental democracy, either citizens as individuals or individuals that have organised themselves as associations.

This book is focused on the power of the individual, since the term environmental democracy reflects the recognition that environmental issues must be addressed by all those affected by their outcome, not just by government. The starting point is the belief that every single person can really act to protect and improve the environment, and the fact that a single person is alone does not mean that s/he is weak *vis-à-vis* the future choice. Every individual has to rediscover what their environmental rights are, s/he exists as a human being and that without their explicit granting, they nevertheless exist.

At the same time, just as with regard to environmental rights, ecological duties exist beyond any recognition. In other words, from the mere fact that we are alive, we have rights and duties vis-à-vis ourselves and Earth.

Such moral and ethical acknowledgements have to be included in the legal concept of the individual, in particular in the notion of citizenship. The new citizenship comprises two aspects: first, environmental citizenship, which entails environmental rights, and second, ecological citizenship, that covers ecologicalduties.

However, both rights and duties may remain unfulfilled "as long as persons do not have the capacity to act in a civil society" (Stec, & Casey-Lefkowitz, 2000, p. 17). Thus, new processes and more participation in decisions, which affect the environment have to be concretely planned and organised by regulation.

Finally, the role of the association will be briefly analysed, as it has been defined as the "blood and sinew" of society since it is "the gate to transforming the industrial state and revivifying community" (Morrison, 1995).

Hence, the actions of individuals and associations could balance the power of the central state and contribute to the shift towards the creation of an environmental democracy.

From this perspective, Section II of Chapter II has been divided into two parts. After an analysis of the theoretical construction of a new citizenship,4 the first part, in particular, will deal with the environmental citizen and his corresponding substantive and procedural environmental rights. The second part will focus on ecological duties corresponding to ecological citizenship.

3 Concrete Solution

As mentioned above, environmental democracy should be implemented at global and local levels to better answer to global and local environmental problems.

In the light of the theoretical construction of environmental democracy and its elements, Chapter II, titled **"Environmental Democracy in an International Context"**, examines environmental democracy at the global level by referring to international legal instruments. An overview of democracy at the local level would have been too extensive for the framework of this book, which focuses only on the global level.

3.1 Global Environmental Democracy

Chapter II explores whether the theoretical construction of environmental democracy, examined in Chapter I, can be found to exist totally or partially on a global level. At this level, the creation of environmental democracy would be achieved through international environmental law, which encompasses in particular treaties the tools by which the international relationship is determined.

---

4 The first Author which spoke about environmental citizenship was M. J. Barker, 1970, p. 33.

It examines the situation and the steps already taken at the international level, in particular, through the Aarhus Convention on Access to Information, Public Participation in Decision-Making and Access to Justice in Environmental Matters.[5]

This treaty has been chosen because it could serve as a model for the democratisation of international environmental decision-making processes and for the construction of an environmental democracy at a global and then at a local level (Marshall, 2006, p. 126; Redgwell, 2007, p. 163). The Convention aims at promoting and developing in a concrete form the environmental democracy and corresponds more closely to the theoretical construction of this new democracy (Wates, 2005b, p. 393). At the same time, this international instrument also recognises the environmental rights and ecological duties of the individual.

Before talking about the substantive and procedural environmental rights and ecological duties in international law and in the Convention, Section I of Chapter II **"Elements of Environmental Democracy at the Global Level"**, intends to outline briefly the conceptual framework of "democracy" and "environment" in international law and in the Aarhus Convention, and additionally intends to give a panoramic view of the steps already taken at the international level towards an environmental democracy.

Section II **"Procedural Environmental Rights in the Aarhus Convention"**, focuses on the provisions of the Aarhus Convention and deals with the three official pillars, Access to Information, Participation and Access to Justice and the two additional pillars, Enforcement of Environmental Law, and the Review of Compliance Mechanism, which are recognised by the treaty.

The aim of this part is not to consider all of the detailed provisions of the Convention but rather to take a broader look at the Aarhus version of the theoretical model of environmental democracy; but to show that the Aarhus pillars represent concrete examples of tools which could help introduce some elements of the theoretical model of environmental democracy primarily at a global level but also at a local level.

It must be pointed out here that several problems and ambiguities in the Aarhus Convention have been identified by the legal doctrine. It has been therefore considered a "fairly weak legal document, given its quite vague and permissive character and the absence of adequate enforcement mechanisms" (Lee, & Abbot, 2003, p. 81). Nevertheless, it can be stated that the Convention makes a *potentially* powerful proclamation with regard to the significance of

---

5 See Convention on Access to Information, Public Participation in Decision- Making and Access to Justice in envirŌonmental Matters, Participants, June 25, 1998, 38 I.L.M. 517 (1999), entered into force Oct. 30, 2001.

public participation in an ample variety of decisions, and it should therefore not be forgotten that the treaty constitutes a "floor, not a ceiling" (Stec, & Casey-Lefkowitz, 2000, p. 45-47).

States have the obligation to provide for broader access to information, more extensive public participation in decision-making and a wider access to justice in environmental matters than required by the Convention. In other words, the Convention sets forth few requirements that parties must meet *at a minimum* in order to provide the basis for global and international environmental democracy, namely the effective recognition of aforementioned procedural rights in environmental matters.

Hence, the ratification process by the states of the Aarhus Convention solves some of the mentioned difficulties.[6] Indeed, it is well known that for international environmental law to be effective, it relies upon its implementation within domestic orders as well as its enforcement.

Furthermore, building environmental democracy has to be reflected at the local level through regional and national regulation. Thus, some of the obligations within the Aarhus Convention, which are considered as *weak*, are likely to be given some real teeth *via* regional legislation and national legislation (Wates, 2005b, p. 393).

---

6 "It is notable that the Aarhus Convention makes no comparable attempt to broaden participation. The real emphasis in the Aarhus Convention is on the involvement of NGOs. However, we should always be aware of the dangers of claiming that NGOs 'represent' anybody, and of the possibility that a small (even if larger than before) number of participants will wrap up important decisions. More generalised public participation of course faces real obstacles". Lee, & Abbot, 2003, p.107-108

CHAPTER I:

Environmental Democracy: a Theoretical Construction

Section I: Environmental Democracy

A CREDO FOR DEMOCRACY

DEFINITION
I Believe that democracy is a positive political process for working toward liberty, equality, and fraternity, and that, though it bears in itself the means of improvement, it can never lay claim to perfection without destroying its essentialnature.
PURPOSE
I believe that democracy seeks to preserve and to reconcile the rival Socrates and moral values o the individual on the one hand and of society on the other, as positive aids toward a higher moral order.
METHOD
I believe that democracy operates by seeking successive compromises in order to maintain a balance among constantly changing alliances of social interests; and that these compromises are expressed in laws which are supreme and can be changed only by the will of the people...
RIGHTS
I Believe that democracy ensures the supremacy of law by guaranteeing to the people certain civil liberties which in substance are not subject to compromise; that among these are freedom of speech, press, religion, and assembly; the right to a day in court; and the right to change their government or its policies by the exercise of the franchise in order to promote the public welfare.
DUTIES
I believe that democracy depends upon the balance wheels of self-restraint and moral courage. Self-restraint teaches the people when to forego their own desires and opinions; it is the basis of social order. Moral courage demands that the people stand up for what they believe is right, whatever the consequences; it is the means by which society advances. These balance wheels cannot function unless the people are taught to know their daily rights and duties and to exercise them faithfully and intelligently; to recognise and prevent the undermining of

civil liberties even at the sacrifice of consistency during a crisis; and to return to
tolerance and compromise when the crisis is past
(Baldwin, 1956, p. XI).

"Since the vote has been extended to every adult citizen, without class, gender,
or racial discrimination, etc., the contemporary challenge of strengthening
democratic regimes is not mainly about who participates, but how, when and
where citizens should participate"
(Bobbio, 1984, p. 44–46).

This first Chapter will develop a theoretical and conceptual framework for the creation of a new form of democracy. Before speaking about the elements, form, space and actors, which compose environmental democracy, it is necessary to analyse what the notions of "democracy" and "environment" in the book's structure encompass.

In order to achieve this, the first section has three main issues, structured into three subsections: the first explores different theoretical forms of democracy, in particular, the participative and deliberative theories, which are both fundamental for the purpose of this book; the second is centred around an analysis of the meaning of the term "environment"; and the aim of the final one is to unify the two analysed notions into one concept, called "environmental democracy", in order to contribute to a construction of a form and spatial dimension of this new type of democracy.

1 "Democracy" and "Environment"

1.1 What is Democracy?

If we are to deal with this question, it must first be clarified that there is no single "democratic theory" but rather there are democratic theories (Dahl, Chicago, 1956, p. 1). Thus, it will be necessary, as a first step, to explore the origin and meaning of this political concept and then go into the essentially contested meanings of democracy.

1.1.1 The Origin of Democracy

Let us begin with a minimal, but generally accepted, definition of democracy as a political system in which the opportunity to participate in decisions is widely shared among all adult citizens (Dahl, 1991, p. 6).

The word "democracy" originates from Greek and literally means "rule by the people", or in other words, the collective power popular sovereignty. The Greeks gave us the word, and also provided a primitive model,[7] where people were capable of taking political decisions by a direct vote on questions. Nevertheless, the Greeks had "little or no idea of the rights of the individual, an idea which is tied up with the modern concept of democracy" (Birch, 1993, p. 45). Indeed, they only granted a small minority of adult inhabitants of the city the right of political participation. This last feature characterised also the *Res Publica* in Roman times when the governance of it was reserved to the patricians, or aristocrats. Only later, the common people also "gained entry". Despite this first improvement, in comparison to the Greek system, the population was still divided between free persons, and slaves, who never received such rights.[8]

From this brief historical introduction, it is clear that the Greek and Roman assumptions and practices were very different from the system of representative government which has developed in the Western World during the past two centuries and which describes the majority of political systems in the world.[9] Now, the notion of democracy is more the *demos'* power to take general political decisions, within a framework of equality and freedom (Arblaster, 1994, p. 8). The definitions given to the modern concept of 'democracy' are so many that they cannot be analysed in detail here.[10] Nevertheless, in the following it will be explored how this notion has been implemented in tree models.[11]

---

7   See in general about the Greek democracy and the system of elections: Hansen, 1991; Accame, 1998, p. 11; Manin, 1997; Staveley, 1972; Finley, 1973; Finley, 1983.

8   For more details see Nicolet, 1978, p. 282.

9   It should be remarked also that a large number of the States today uses this model. R. Dahl has affirmed: "If we examine the best known example of Greek democracy, that of Athens, we thus notice an important difference from our present version. A political institution that the Greeks saw not only as unnecessary for their democracies but downright undesirable was the election of representatives with the authority to enact laws. We might say that the political system they created was a primary democracy, an assembly democracy. But they did not create representative democracy as we understand it today". Dahl, 1998, p. 7.

10   Indeed some authors have given 311 definitions of democracy. Naess, 1956.

11   Bobbio's definition of democracy. The author defines 'democracy' as a procedural democracy (democrazia procedurale), as a whole of game rules, which provide for: 1) who is authorized to take the decisions (competence); 2) the modalities through which decisions are taken (proceedings): see Bobbio, 1984, p. 4.

1.1.2 Different Models of Democracy

1.1.2.1 Representative Democracy

Representative democracy is a modern form of democracy, first expressed in the late 18th century in the founding statements of the American and French Republics where "it emerged as the moral justification for the lawful authority of the State over large populations" (Mason, 1999, p. 21). In the modern sense, the term democracy describes a system of representative government in which the representatives are chosen in free elections.

As Birch has underlined, "the opportunity to vote is the minimum condition that a governmental system must satisfy to qualify as democratic, but further opportunities and forms of political participation are highly desirable" (Birch, 1993, p. 46). In fact, new claims of broader democratic spaces and more measures for involvement of citizens in political life have been solicited by civil society.

Representative democracy seems to be insufficient today (Rodriquez, 2008, p. 37), since the serious problem is that this form of democracy suffers from a lack of legitimacy, also called "democratic deficit". This takes its form in the distance between the representatives and those represented,[12] in the failing action of the parties, as well as in a lack of publicity, accountability and transparency at higher levels (Habermas, 1996; Vitale, 2006, p. 748).

To solve such a democratic deficit, new ways have appeared in the shape of participatory and deliberative democracy. Under both models, participation of the public is functionally and morally central to democracy (Webler, & Renn, 1995). The observation of political scientist literature during the late 1960s and the early 1970s already show an emerging trend of themes related to participation. These trends "initiated an enhancement of the typology for democratic regimes or at the least, theories of democracy, this is due largely to the fact that during and since that late 1960s and early 1970s, participatory democracy has been contrasted to representative democracy or other versions

---

12   On the matter, see Rodriquez, 2008, p. 37: "For many years, representation appeared to be founded on a powerful and stable relationship of trust between voters and political parties, with the vast majority of voters identifying themselves with, and remaining loyal to, a particular party. Today, however, more and more people change the way they vote from one election to the next, and opinion surveys show an increasing number of those who refuse to identify with an existing party. Differences between the parties once appeared to be a reflection of social cleavages. Each party used to propose to the electorate a detailed program of measures which it promised to implement if returned to power. Today, the electoral strategies of candidates and parties are based instead on the construction of vague images, prominently featuring the personality of the leaders". See also Mersel, 2006, p. 84.

of democratic elitism". After a long period during which the concept of participation had lost most of its relevance in public and scientific debates, a new concept, namely participatory and deliberative democracy, has become a prominent topic in many publications.[13]

1.1.2.2 Participative Democracy

Jean-Jacques Rousseau did not believe in a representative government, due to the fact that men shall not be represented by others (Rousseau, 1922).[14] The main keys to his thinking were the idea of direct self-government in small communities, and that "individuals should put their personal interests aside when they participate in politics, and commit themselves instead to the promotion of the communal welfare".[15]

Indeed, if the law of the State were supported by the general will, it would not restrict the liberty of citizens and it forces only obedience to those laws which one had consented to in the first place. Hence, to reach this model the community must be small enough for its citizens to assemble and express directly their votes, without representatives (Rousseau, 1913, p. 121).

Ernest Barker of Cambridge followed, with reference to some aspects, the same tendency, when he wrote that the real basis of democracy is "the discussion of competing ideas, leading to a compromise in which all the ideas are reconciled and which can be accepted by all because it bears the imprint of the all" (E. Barker, 1942, p. 41).

The conception of participatory democracy, as the term suggests, considers *participation* to be the fundamental feature of political practice. Some authors defined participatory democracy in simple literal terms, as 'rule by the people', as "all acts of citizens that are intended to influence the behaviour of those empowered to make the decisions" (Chekki, 1979). As a concept, participation refers to "a general idea of inclusion, equality, representatively, legitimacy,

---

13 See for example the European Commission's White Paper on European governance 2001. See on the matter: Greven, 2007, p. 233.

14 See also, on the matter: Fralin, 1978.

15 Rousseau believes that the citizens could have two levels of consciousness: on the one hand they would be conscious of their own individual or group interests, leading to a see of 'particular wills' to promote measures favourable to those interests. On the other hand, they could be led to think in terms of the interests of the community as a whole, leading to a real will to protect measure that would protect these shared interests. The first category of wills of citizens would be diverse and to some extent mutually incompatible, the second category of the wills would merge into a consensus that Rousseau called the "general will". See Birch, 1993.

voice. It can be defined as a process through which individual or collective actors influence and share control over the decisions and resources that affect them".[16]

Participation has to be understood as encompassing also "a complex spectrum of activities", which may be divided into the political arena and the administrative arena. In other words, on the one hand, participation in policy-making processes exists, called 'political participation'; on the other hand, participation in administrative proceedings, called "participation in administration" exists (Rodriquez, 2008, p. 24).

Political participation is "public involvement in expressing preferences for a broad spectrum of important [...] policies, mainly during the process of selecting political representatives, campaigning and voting" (Wang & Wart, 2007, p. 265). Going deeper, the definition of political participation means "participation in the process of government"[17] and occurs mainly at the legislative levels (Wang & Wart, 2007, p. 265).

Participation in administration is understood to be "public involvement in administrative process and administrative decision-making [...], whereas participation in administration is realised at the executive level" (Wang & Wart, 2007, p. 265).

Hence, participatory democracy may embody the aforementioned declinations of participation and the introduction and the use of this model can be justified on three bases. Importantly, citizenship is reconstructed and improved because political practice is enlarged, even beyond the representative system (Barber, 1984, p. 154).

Then, the amplification of participation can directly legitimise the decision-making procedures of the state and improve the quality of decision-making (MacPherson, 1977, p. 94). In fact, participation enhances "the opportunities for mutual accommodation through exchanges of reasoned arguments" and generates "higher levels of trust among those who participate" and finally "this, in turn, allows them to introduce a longer time-horizon into their calculations

---

16 From the World Bank's definition of participation in development: "a process through which stakeholders influence and share control over development initiatives and the decisions and resources which affect them" World Bank, 1996, p. XI.

17 The principles forms of political participation can easily be listed, and are as follows: voting in local or national elections, voting in referendums; canvassing or otherwise campaigning in elections; active membership of a political party; active membership of a pressure group; taking part in political demonstrations [...]; various forms of civil disobedience [...]; membership of government advisory committees; Membership of consumers' councils for publicly owned industries; client involvement in the implementation of social policies; various forms of community action; such as those concerned with housing or environmental issues in the locality". Birch, 1993, p. 80-81.

since sacrifices and losses in the present can be more reliably recuperated in future decisions" (Heinelt, 2007, p. 220).

A final justification to introduce this model is, according to some traditions in democratic theory, that participation produces better results (Schmalz-Bruns, 2002, p. 59). This is argued in several points. First of all, one basic normative assumption of democratic theory, starting from the idea of the natural rights of individuals, is that "those who are affected by a decision shall be given rights to participate in the process leading to the outcome of that particular decision" (Heinelt, 2007, p. 220).

Thus, even if the final decision is not based on their will, they have had the opportunity to make their arguments heard. Secondly, participation by individuals with a broad range of interests, allows all participants to offer reasons for their position which aids "in the elimination of egoistic and illogical positions" (Gbikpi, & Grote, 2002, p. 17; Pateman, 1970).

Theoretical participatory theory aims at transforming the model of "thin democracy" (Barber, 1984, p. 14), as existent in practice and mostly limited to representatives, into a "strong democracy", which is to be exercised and benefited from by citizens, who can participate in arenas other than voting.

Thus, from a practical and organisational approach, participatory democracy emphasises the construction need of forms of direct democracy, which can function in conjunction with the representative system.

Following Bobbio's view, representative democracy and direct democracy are not alternative systems, but systems which can complement each other,[18] because the principle of popular sovereignty is the common ideological and historical basis of both representative and direct democracy (Rodriquez, 2008, p. 37). Therefore, while it is clear that a *tabula rasa* of the indirect with the direct system is not the intention, the aim nevertheless is to create new spheres of discussion and political deliberation, which remove or at least diminish the grave problems of legitimacy (Vitale, 2006, p. 748).

A last point related to this model is the involvement of the public through the inclusion of NGOs. Indeed, these organisations promote citizens' participation and they have a significant impact not only on the political activity of the country or region involved,[19] but also on civic involvement in matters affecting

---

18 Bobbio: "Democrazia rappresentativa e democrazia diretta", 1978, p. 22, "Democrazia rappresentativa e democrazia diretta non sono due sistemi alternativi, nel senso che laddove c'è l'una non ci può essere l'altra, ma sono due sistemi che possono integrarsi a vicenda". On this point, see also Rensi, 1995.
19 Bacqué, Rey, & Sintomer, 2005, p. 10: "Parallèlement, à l'échelle internationale, des mouvements sociaux luttant contre la mondialisation néolibérale se sont affirmés, souvent coordonnés en réseaux peu hiérarchiques, tandis que les ONG jouent un rôle croissant

citizens' lives" (Rodriquez, 2008, p. 24), in particular relevance for this book, in the environmental protection field.

Concerning this relatively new phenomenon, which will be expanded later on, it must be underlined that for the purpose of this book, the more important point is the role played by citizens and not by NGOs which have an accessory role. In general, attention will be laid on individuals, even if the NGOs are discussed to some extent as well, especially since in most legislation more rights are recognised to the civil organisation than to the singular citizen.

1.1.2.3 Deliberative Democracy

The deliberative theory of democracy recognises its point of departure, that currently a 'democratic deficit' characterises all democracies in Western countries and suggests a new approach to the new reality (Ostrogorski, 1922, p. 55).[20] Deliberative democracy has been described as "the practice of public reasoning," in which "participants make proposals, attempt to persuade others, and determine the best outcomes and policies based on the arguments and reasons fleshed out in public discourse" (Schlosberg, Shulman, & Zavetosk, 2006, p. 216). The distinguishing element of deliberation is an open discussion, in which participants are given equal treatment, respect and opportunities (Saward, 2001, p. 564).[21]

Many authors,[22] thus, support their theories with Habermas' notions of deliberation, in which "deliberation refers to an attitude toward social cooperation, that of openness to persuasion by reasons referring to the claims

---

et qu'elles commencent à être associées au moins à la marge aux cercles de décision. Partout, les modes traditionnels de gestion et d'administration sont remis en cause".

20   On the matter, see Schmitt, 1998, in part. p. 90: "[...] Se la situazione del parlamentarismo è oggigiorno così critica, è perché l'evoluzione della moderna democrazia di massa ha fatto della discussione pubblica, con i suoi argomenti, una vuota formalità [...]. I partiti [...] non si affrontano più oggi sul piano delle opinioni da discutere ma, come gruppi di gruppi di pressione sociali o economici, essi valutano i loro interessi e le loro rispettive possibilità di accesso al potere e, su questa base fattuale, concludono compromessi e coalizioni. Le masse vengono conquistate grazie ad un apparato di propaganda [...]. Il ragionamento, quello che è caratteristico della discussione, è destinato a scomparire. Al suo posto si ha, nei negoziati di parte, il calcolo ben ponderato degli interessi e delle probabilità di accedere al potere".

21   Schmalz-Bruns affirms: "Deliberative understanding of democracy suggest a political practice of argumentation and reason giving among free and equal citizens, a practice in which individual and collective perspectives and positions are subject to change through deliberation and in which only those norms, rules or decisions which result from some form of reason-based agreement among the citizens are accepted as legitimate"; Schmalz-Bruns, 2007, p. 283.

22   See also Bessette, 1980, p. 102; Cohen, 1989, p. 17; Nino, 1996.

of others as well as one's own" (Habermas, 1998, p. 244).[23]Habermas moreover constructed the concept of democracy from a procedural dimension (Habermas, 1987b, p. 163; 1987a, p. 340; 1996, p. 177). Thus, to achieve democratic legitimacy, it is required that the process of political decision-making occurs within a framework of broad public discussion, in which it is possible for all participants to discuss the different issues in a watchful and rational manner. Decisions can be prepared only after this method of debate has taken place.

Democratic deliberation is best explained as "being orientated towards a mutual understanding, which does not mean that people will always agree, but rather that they are motivated to resolve conflicts via arguments rather than by other means" (Graham, 2003, p. 59; Warren, 1995, p. 181).

Moreover, deliberative democracy implicitly implies an active notion of citizenship (Graham, 2003, p. 59).[24] In fact, by acknowledging the citizens as the main players in the political procedure, political deliberation involves a strong model of participation. Consequently, Cohen elucidated the concept of deliberative politics in terms of an "ideal procedure" of deliberation as well as decision-making, which should be "mirrored" in social institutions as much as possible (Cohen, 1989, p. 17; Habermas, 1996).

Indeed, promoters of deliberative democracy are also conscious of the relationship between social and economic rights and political equality. Habermas underlined that those procedures have to be set up by law:[25] "the processes and conditions for the process of democratic opinion- and will-formation are institutionalised through the *medium* of law, crystallising in a group of fundamental rights".[26]

Furthermore, Habermas acknowledges the need to implement social, economic and environmental rights since they are vital for the enjoyment of the rights of communication and participation.

Moreover, according to the majority of scholars, deliberative democracy demands a level of social and political equality. It is important to note that

---

23   Habermas emphases communicative action entails using knowledge in speech to convince others of the validity of claims Prerequisites for communicative action, and thus for deliberative democracy, would thus include rough equality, educational competence, and shared cultural and linguistic understandings. A distinctly different approach to deliberative democratic theorising can be found in the work of John Rawls and others building on that tradition.

24   As Steele remarks, "Citizens become deliberators", Steele, 2001, p. 415.

25   See foot note 18 in Habermas, 1996.

26   He claims that "The legal system as a whole needs to be anchored in basic principles of legitimisation. In the bourgeois constitutional State these are, in the first place, basic rights and the principle of popular sovereignty", in Habermas, 1987b, p. 178; Habermas, 2006, p. 748.

deliberative democratic proponents do not explicitly acknowledge identical prerequisites (Habermas, 1998, p. 244).[27] The difficult requirements of deliberation have been remarked for a long time. For example, Aristotle highlighted that "sameness" and equality were necessary conditions (Aristotele, 1946; Bohman, 1996, p. 109).

Rousseau revealed social and economic equality and cultural homogeneity as prerequisites for self-rule as well. A complete examination of prerequisites for deliberation is neither possible here, nor necessary. Nevertheless, it is possible to summarise the basic conditions mentioned for deliberative democracy: socio-economic and political equality, education or literacy, cultural homogeneity, a level of overall societal wealth, the social and cultural norms of modernity and pluralism.

If all the indicated conditions were in fact necessary for the successful implementation of deliberative democracy, then there would not be a high probability that deliberative democracy could be established in societies which are poor, predominately illiterate, and culturally heterogeneous or any combination thereof. However, there have been a number of examples of deliberative democracy which can be found in places where some or all of these conditions did not hold. This proves that these preconditions in fact are not essential after all, and that the deliberative democratic theory should be implemented in all the more or less traditional societies (Gupte, & Bartlett, 2007, p. 94).[28]

---

27   Habermas emphases communicative action entails using knowledge in speech to convince others of the validity of claims Prerequisites for communicative action, and thus for deliberative democracy, would thus include rough equality, educational competence, and shared cultural and linguistic understandings. A distinctly different approach to deliberative democratic theorising can be found in the work of John Rawls and others building on that tradition (Rawls, 1999) Rawls argues that such conditions include decencies in political traditions, law, and property; class structure, religious and moral beliefs; culture; human capital and know-how; material and technological resources; and enough wealth to realise and preserve just institutions. Amy Guttmann and Dennis Thompson (Guttman, & Thompson, 1996, p. 358), who argue for a "full liberalism" approach to deliberative democracy identify literacy and numeracy as "prerequisites for deliberating about public problems". Deliberative democracy can thereby further devalue already marginalised groups. Some detractors argue that the deliberative model assumes cultural neutrality and universality and does not acknowledge that power enters speech Young (Young, 2000, p. 123). On the other hand, proponents of deliberative democracy believe that an important function of deliberation is to discipline the exercise of power through the common reason of citizens. Cohen and Rogers discuss three conditions for empowered participatory governance—focused problem-solving, participation, and deliberation "The deliberative ideal of using common reason to discipline power and preference thus arguably connects to substantive norms of political equality (fairness of procedure) and distributive equity (fairness of result)". Cohen, & Rogers, 2003, p. 242.

28   In this article the Author examines a case of village deliberative democracy in a

1.1.3 Conclusions about Democracy

The point of departure in participatory, as well as deliberative theories, is the recognition of the fact that there is a crisis of political legitimacy, which must be overcome. In both theories, the return of legitimacy lies in the need of a more participatory political structure, and the improvement of democracy takes place in a continuous and dynamic process of democratisation *of democracy*. This further modifies democracy progressively into a more comprehensive system, which concerns "the constitution of societies and emancipated forms of life".[29]
The participatory debate is focused on the necessity of employing procedures of direct democracy and on the importance of expanding these procedures to encourage substantial democracy in a way that reduces social and economic disparities and assures the successful enjoyment of political rights for individual. The focal point of deliberative democracy is "the exercise of sovereignty in the discursive processes of collective will-formation, which must be legally institutionalised" (Vitale, 2006, p. 748). "Deliberative theory, recognising the informal space of public opinion as essential for the political process of discursive development, defends the expansion of spaces in which will-formation constitutes itself" (Vitale, 2006, p. 759). The solution, also common to the aforementioned conceptions, is to re-absorb citizens in the public debate and political procedures by means of participation and public deliberation.

Thus, the integration between the two models offers a better option for the enhancement of democracy because the two models are compatible and complementary. In other words an integrated democratic system, which unites mechanisms of direct participation and deliberation with instruments of representative democracy,[30] should diminish the democratic deficit.

To sum up, it is necessary to underline two points: first, the deficit in the actual model of representative democracy and second, a movement towards an emerging new solution derived from participatory and deliberative models of democracy. Both are important for this book because they are the form on which environmental democracy is based, as will be discussed below

---

developing country against these basic assumptions of deliberative democracy theory. One village within the Indian State of Maharashtra was examined with regard to its community conservation processes. This village was part of a larger research project that examined community conservation in two States and four villages in India.

29   This is the phrase Habermas uses in his preface to Between Facts and Norms.

30   As the system of political parties, the parliament and the executive power.

1.2 What is Environment?

> The most famous definition of "environment" is that given by Albert Einstein
> who once said "*The environment is everything that isn't me*"

This second question is not susceptible to one generally accepted answer, since to understand the significance of the term "environment"; it is desirable to evaluate the position of "man" in relation to the natural order. Consequently, environment and the degree of implementation of the "environmental democracy" are conditioned on the relationship that man chooses to have with his surrounding ecology. This section briefly introduces the origin of the term *environment* and a number of definitions of environment, stemming both from an anthropocentric and ecocentric vision of nature.

1.2.1 The Origin of the Term Environment

The origin of the term environment is French, deriving from "environner", literally meaning "to encircle". Since the beginning of the 1960s, new words have become apparent in several languages to convey the notion of environment: "Umwelt" (German), Milieu (Dutch), Ambiente (Italian), Medio Ambiente (Spanish), Meio ambiente (Portuguese). The word "environment" emerges from the concern of potential damages to natural resources and "the processes on which life depends" (Kiss, & Shelton, 2000b).

According to the Webster's Dictionary, the general definition of environment is "the circumstances, objects, or conditions by which one is surrounded" (Webster's Ninth New Collegiate Dictionary, 1983). It continues with a more precise explanation: "the complex of physical, chemical, and biotic factors (such as climate, oil, and living things) that act upon an organism or an ecological community and ultimately determine its form and survival" and "the aggregate of social and cultural conditions that influence the life of an individual or community". The last definition is very wide and covers "urban problems such as traffic congestion, crime and noise within the field of environmental protection" (Kiss, & Shelton, 2000b).

Going deeper into the issue of establishing a definition of the term environment, the question arises whether the environment has to be defined in relation to humans, as a requirement for enjoyment of other goods (life, health, property, quality of life) or as an autonomous good.

Different economic paradigms underpin both relations. The first paradigm presupposes that everything on earth is for the sole use of humankind and that our species is at liberty to modify the environment at its will. The value of the environment is determined "by economic rationality as a monetary process reflecting market force of supply and demand" (Hancock, 2003, p. 23). The environment is therefore essentially perceived as a good within this paradigm. Consequently, there is a separation of human society from ecological systems, and the environment is only valued "within a framework of economic rationality to the extent that the market mechanism specifies prices for natural resources" (Hancock, 2003, p. 23).

In contrast, the second paradigm attributes intrinsic value to non-human life, independent of its economic or anthropocentric worth. This sentiment has been articulated through the claim that "everything has some value for itself, for others and for the whole" (Drengson, 1998, p. 221; Hancock, 2003, p. 23).

The two economical concepts of the environment correspond indeed to the two contrasting views that individuals and societies hold concerning the relationship between human species and the natural order (Cooper, & Palmer, 1998): that is, anthropocentric view and ecocentric view.

Nature has always exercised a mysterious fascination for man. Primitive man regarded the elemental forces of nature with awe and respect, and identified them as deities to be feared and propitiated. Thus, as it will be seen, in some early civilisations, ancient cultures and religions drew their values from that relationship. Despite these roots, man quickly began to modify this view and to move towards the idea that man is the reason for all creation (Passmore, 1976, p. 17).

In fact, man, little by little, has continued in a "state of alienation from nature" abusing and degrading the planet's ecological system. The socio-economic pressure on natural resources has reached the point where the quality and condition of human life is threatened and has put into question the very survival of the human race. During the latter half of this century, the enormous power provided by the advanced sciences and high technology has given an impetus and a momentum to environmental problems that enable them to influence living conditions in distantly separated territories" (Pathak, 1992, p. 205-206). Man, who personifies the reflection and the image of God, believes that all the natural elements are at his feet. By following this approach, the environment "is nothing other than an element submitted to man and to his necessities" (Fraccia, 2009).

This anthropocentric approach has been consolidated by modern man's ability to manipulate nature, especially through technology. Even the idea of endless and unlimited progress, made possible by science, has led to such domination that nature is viewed as an element in human hands. Environmental degradation is tightly linked to the shift towards the anthropocentric interpretation of the relationship with "nature", so, in economic terms, to identify the environment as a good among other goods.

1.2.2 Different Approaches to the Environment

1.2.2.1 Anthropocentric Approach

As seen above, it must be considered that the main notion of environment is based in anthropocentrism. The basis of this was created through the religious concept of the human person, which was considered to be the centre of the universe (Pace, 2001, p. 15). This idea was in particular disseminated by the Christian thought (Zamagni, 1994, p. 235-237).

The attitude was that "since everything is for men, he is at liberty to modify it as he will" (Passmore, 1976, p. 17). This belief betrayed man into the false assumption of superiority over the natural order. He misunderstood the role of stewardship of the planet as an "absolute proprietorship" (Passmore, 1976, p. 17). In the dynamics of daily existence, human life has been lived in the dimensions of an anthropocentric perception "that treats the rest of Creation as bonded in subservience to it".

Genesis is the starting-point to better understand this attitude. The Lord God created man, so Genesis certainly narrates, to have "dominion over the fish of the sea and over the fowl of the air, and over the cattle, and over all the earth and over every creeping thing that creep upon the earth" (I:26). This has been read not only by Judaism, but also by Christianity and Islamism as man's charter, granting him the right to subdue the earth and all its inhabitants. Moreover, God, according to Genesis, also issued a mandate to "be fruitful and multiply and replenish the earth and subdue it" (I: 28). Thus, Genesis tells men not only what they can do, but what they should do to multiply and replenish and subdue theearth.

This passage is also repeated after the Flood: God still exhorted Noah to "be fruitful and multiply and replenish the earth" but then he added two significant points. The first made it clear that men should not expect to subdue the earth either by love or by the exercise of natural authority, as distinct from force: "and the fear of you and the dread of you shall be upon every beast of the earth, and upon every fowl of the air upon all that move upon the earth and upon all the fishes of the sea: into your hand are they delivered". The second, "every moving thing that lives shall be food for you", permitted men to eat the flesh of animals.

Although the Old Testament insists on man's domination, it is far from suggesting that God has left the fate of animals entirely in man's hands, whether before or after the Fall (Passmore, 1976, p. 7).

There are two possible interpretations of the above quotations: the first, man is nature's absolute master, for whom everything that exists was designed. Following this view, since everything on earth is for man's use, he is at liberty to modify it, as he will.

The second interpretation of this passage of the Bible is that "he takes care of the living things over which he rules for their own sake, governing them not 'with force and with cruelty' but in the manner of a good shepherd, anxious to preserve them in the best possible condition for his master, in whose hands alone their final fate will rest" (Passmore, 1976, p. 9). Thus, one can speak of "Christian arrogance" and this approach was long predominant and did also not find an obstacle in the modern scientific vision of nature.

In the seventeenth century, Bacon said indeed "the empire of man over thing depends wholly on the arts and sciences", man seeks to gain intellectual knowledge of nature, overcoming her resistance not by force but by his intimate knowledge of her secrets. Man could become not only the titular, but the actual lord of nature, because knowing nature also entails restoring it through science. In Bacon's book *New Atlantis* he said: "the end of our foundation is the knowledge of cause, and secret mooting of things; and the enlarging of the bounds of human empire, to the effecting of all things possible".

Descartes also agreed with Bacon's view. He aspired to a "practical philosophy by means of which, knowing the force and the action of fire, water, the stars, heavens, and all the other bodies that environ us, as distinctly as we know the different crafts of our artisans, we can in the same way employ them in all those uses to which they are adapted, and thus refer ourselves the masters and possessors of nature" (Descartes, 1931). Descartes' emphasis is on the scientist-technologist, who freely makes use of a nature which was not created only to serve him but which is so constituted as to be potentially useful to him.

Moreover, in his view every finite existence except the human mind is a mere machine which men, in virtue of that fact, can manipulate without scruples. No doubt, Descartes has taken from the Christian tradition towards the attitude of man, which thinks of him as nature's governor. The ideal of mastery could thus persist in Europe even when the Bible had lost much of its old authority.

This approach was later maintained and reinforced also in the nineteenth century. Spencer transformed Darwin's theory of natural selection into the doctrine of the 'survival of the fittest'. Man, it was alleged, not only had to struggle against nature in order to survive, but demonstrated his moral superiority by his success in doing so (Passmore, 1976, p. 23).

Furthermore, Bacon and Descartes interpretations were absorbed into the ideology of modern Western societies, communist as well as capitalist, and has been exported to the East. Nowadays, this viewpoint can no longer be sustained; it is no longer possible to maintain the idea of the environment as something to be exploited and submitted by man and his boundless will.[31]

---

31   The historian Simon Shama expresses the point well in his book, *Landscape and*

A second interpretation of Genesis "tells man that he is, or has the right to be, master of the earth and all it contains". But at the same time "it insists that the world was good before man was created, and that it exists to glorify God rather that to serve man" (Passmore, 1976, p. 27).

Hence this second interpretation has recently come into favour.

1.2.2.2 Ecocentric Approach

The anthropocentric approach to nature has been criticised.[32] Marsh's *Man and Nature* (Marsh, 1864) was the first work to describe in detail man's destructiveness, a destructiveness arising out of his 'ignorant disregard of the laws of nature'. In order to civilise the world, man was forced to transform the relationship between the elements of nature; however, where man went wrong was in supposing that he could act thus with impunity.

The author affirms that: "the ravages committed by man subvert the relations and destroy the balance which nature had established between her organised and her inorganic creations". What he saw is that nature is not a passive recipient of human action and men have recently been reminded in a variety of unpleasant ways, e.g. hurricanes, climate change, etc.

Thus, this ecocentric approach which originally comes from the first and ancestral relations between nature and man, and which is still present in some religious and philosophical views around the world, presents a different concept of the "environment", one in which all organic existence in a single framework are united in harmonious interactions.[33]

In Taylor's opinion the most accurate summation of the ecological model is embodied in the statement that "the universe must be seen as a systematic hierarchy of organised complexity – a myriad number of wholes within the wholes, all of which are interconnected and interacting. Within this perspective, an individual system cannot be properly understood apart from its relationship

---

memory, although it "accustomed to separate nature and human perception into two realms, they are, in fact, indivisible. Before it can ever be a repose for the senses, landscape is the work of the mind. Its scenery is built up as much form strata of memory as from layers of rock". Shama, 1995.

32 See Tribe, 1974, p. 1315; Fox, 1989, p. 5; Naess, 1989; Passmore, 1976.

33 "More than 3.000 years ago, the Upanishads in India expressed the Vedantic viewpoint that the Supreme Reality was the undivided whole, the Brahman, which incorporated all manifestations of matter and energy together in a primordial transcendent, all-pervasive and all-binding harmony. In American Indian culture [...] the concept of an ultimate wholeness of all existence was implied when the individual sought relatedness to all manifestations of the Great Spirit: rocks, trees, animals, or people. The Chinese philosophers believed that man must invariably be seen as inseparable from nature and in oneness with the universe": Pathak, 1992, p. 205-206.

with the environment of which it is an integral part" (P. W. Taylor, 1986; Guha, 1989, p. 71; Wilkinson, 2002, p. 228).

So, this approach denies that men, in relation to the environment, is essentially a despot, but it sees him as a 'steward' actively responsible as God's deputy for the care of the world. Thus the Hebrew word "subdue" has to be translated as "men hold their dominion over all nature as stewards and trustees for God [...] they are confronted by an inalienable duty towards and concern for their total environment, present and future; and this duty towards the environment does not merely include their fellow-men, but all nature and all life" (Montefiore, 1970, p. 55; Passmore, 1976, p. 29).

This approach recognises a balance between a right to environment and a duty to conserve species and environment on the grounds that species and the elements of environment possess value (Rolson, 1988, p. 143). Aldo Leopold (Leopold, 1949) summarises this view telling "the role of *Homo Sapiens* from conqueror of the land community to plain member and citizens of it [...] and implies respect for his fellow members, and also for the community as such". Callicott (Callicott, 1980, p. 311) and Devall and Sessions (Devall, & Sessions, 1984, p. 296) also make the point that our obligations are not only to individual animals but also indeed principally to species and the biotic community as a whole.

1.2.2.3 Environmental Law Approach

The above mentioned ecocentric interpretation involves a new environmental conscience and a global consensus as regards the environmental obligation of protection of the Earth, as well as with regard to future generations. Indeed, the growth of environmental problems and environmental degradation forces the citizens of the world to realise that the world is one.

In fact, today, mankind has the "power to change our global environmental irreversibly, with profoundly damaging effects on the robustness and integrity of the planet and the heritage that we pass to future generations" (Weiss, 1990, p. 198). The consequence of this moral dimension has demonstrated that the environment has gained importance among many green philosophies as well as green thoughts by the dimension of duties and responsibilities of States and citizens.[34]

Hence, the legal answer to this situation is the creation of environmental law, functioning as a set of rules to re-establish the equilibrium between man and nature. Environmental law "springs from the understanding that the

---

34   This point will be better analyse in the next section. Attfield, 1983.

environment determines the form and survival of an organism or community; thus, national, regional and international efforts must be taken to ensure the continued viability of the planet and the sustainability of its myriad species" (Kiss, & Shelton, 2000b).

Nevertheless, this answer is not sufficient either, because it is still based on the anthropocentric and not on an ecocentric approach. The idea that *what is good for humans is good for nature* may not be true, or may only be true in certain cases. Thus, following the first view, the protection of environment also has to be achieved for the welfare of the human and not in a holistic approach.

Despite anthropocentrism being the dominant ethic in current environmental law and policy,[35] some authors have tried to justify it, affirming that "anthropocentrism in law may appear inescapable; law is a human institution, after all, primarily designed to advance human needs. Simply by conceiving of an environment and considering how human beings ought to behave in that environment, and positing laws to regulate the same, we are engaging in activities that only human beings can engage in; by nature we approach environmental concerns in this uniquely normative way" (Donnelly, & Bishop, 2007, p. 89).

Nevertheless, man cannot survive without the environment and so he needs to recognise that men 'form a community' with plants, animals, and the biosphere and that every member of that community and the environment itself have intrinsic value and consequence (Passmore, 1975, p. 251).

Thus, the complexities of the 'environment' are best captured by the term "ecological integrity": it reflects the view that there are natural processes necessary to maintain Earth's life support systems that humans and all life depend on. In other words, it is not the environment, but "the interactions between the various life forms – including human beings – we should be concerned with". This definition acknowledges "not only the complexities of the natural world, but also the fact that humans are part of it" (Bosselmann, 2008).

---

35 For example, the concept of intergenerational equity as enshrined by the principle of sustainable development stresses the requirement to fulfill the needs of the current human population without jeopardising the ability of future human generations to fulfill their needs. Also concepts such as "Best Practicable Means", "Best Available Technique Not Entailing Excessive Cost", and "Best Available Technique with an Element of Proportionality" have clear anthropocentric connotations. Those subject to regulation will only be expected to protect the environment insofar as this can be achieved without causing excessive damage to human-centred economic and societal concerns. Traditional natural law suggests, in contrast, that there may be a rationale to advance environmental protection and to cause minimal environmental damage, not with disregard to our interests, but, nonetheless to the maximum of our capacities. There would follow an obligation actively to seek to improve the practicability of the means, to invest in the advancement of available techniques and to facilitate those conditions that are more likely to make it proportionate to assist our environment. (See pp. 8-12).

In conclusion, the focus of environmental law can no longer be entirely anthropocentric: a shift towards an ecocentric approach, which takes into account the interests of human beings, individual non-humans and the environment as a whole is necessary. In fact, a kind of polycentric approach to environmental protection could not only emphasise the convergence of interests between humankind and the environment, but also ensure a "balancing of those interests where they conflict" (Waite, 2007, p. 395).

2 What is Environmental Democracy?

"It was the environmental movement in the North that first challenged the overarching claims to legitimacy by political systems based on representative democracy [...] During the 1970s and 1980s, thousands of young men and women protested against their democratically elected governments on the sitting of nuclear power stations. Simply because certain people had been elected by majority, they argued, was not enough to give them the untrammelled right to decide how the local environment could be used without the consent of the people who were worst affected by this decision. They essentially wanted a deepening of the democratic process"
(Hay, 2002).

After the analysis of the significance of the term democracy and environment, now the aim is to link these notions into one concept and assess which path could be undertaken to achieve environmental democracy. This section will focus, first, on the relationship between the terms and then, on the theoretical construction of environmental democracy as it has been identified by green political thought.

The birth of modern environmental theories are related instances in the decade following 1962: the publication of *Silent Spring*, the "clarion call" on pesticides poisoning from Rachel Carson (Carson, 1962), and the 1972 Stockholm Conference on the Human Environment, which created the basis of the United Nations Environment Programme. In that period, it can be said that the environmental movement and its underlying philosophies were becoming a global phenomenon.[36]

The key ideological distinction within environmental theories concerns their method of construing a State and how it addresses its environmental problems.

---

36 Much work has explored the relationship between democracy and ecology: Eckersley, 2004; Mason, 1999; Howard, 1996; Eckersley, 2001, p. 52; Jasanoff, 1996, p. 2.

The starting point of all green political theories, in all their organisational and ideological diversity, concerns proposals of an alternative to the liberal-democratic representative system,[37] which was recognised early as unable to resolve environmental problems (Graham, 2003, p. 53). Moreover, it should be underlined that ecological thought does not necessarily lead to democracy, and democratic theory is not automatically pro-ecological. Indeed, in the 1970s and 1980s the dominant thoughts were, on the one hand, the authoritarian model and, on the other hand, the anarchical model. Both models shared, despite their radical opposition to one another, a perception of ecological crisis and the conviction that this crisis called for a rapid and dramatic transformation of the established order (Torgerson, 2008, p. 18).

Nevertheless, since the 1990s different theories came up, which, while always sceptical of the greening capacity of liberal-democratic institutions, have envisaged strong democratic alternatives arising, through evolution and not revolution, from existing liberal democratic regimes.

Hence the first part of this sub-section is a brief survey of the key environmentalist positions regarding the above mentioned models, and the second part examines both why the link between environment and form of State has to be democratic and the literature on democracy, in particular with regard to the participatory approach as well as the deliberative approach.

2.1 Different Political Green Theories of Governance to Solve the Ecological Crisis

Political green thought suggests principally three different forms of the State, *e.g.* the authoritarian State, the anarchical State and the democratic State. With regard to this book, it is worth noting that the issue is not only linked to the question of the classic idea of the State but the scope also extends to the international or regional levels.

In this approach it is therefore better to speak about different forms of governance, even if scholars tend to concentrate their different proposals on the nation-state. Thus, not only which scholarly visions could provide for the best green State, but also how these different theories could be adapted to the other levels of governance will be analysed.

---

37 There are two distinguishable aspects: democracy is about registering public values in the overall political decision-making process; liberalism is the notion of refraining from the imposition of view of the good life or right behaviour, especially in the so-called private sphere of life.

2.1.1 Authoritarian Perspective

A generation ago, William Ophuls pronounced "democracy dead". Quoting the growing pressures of ecological scarcity, he declared: "The golden age of individualism, liberty, and democracy is all but over" (Ophuls, 1977, p. 145). According to him, "democracy had gone wrong at two levels. First, it had assumed that rational self-interest was all there was to human beings. Second, it had assumed that interest-group liberalism was all there was to democratic politics. But the relentless and impersonal constraints of the environment were proving this assumption to be, if not mistaken, at least unsustainable" (Baber, & Bartlett, 2005, p. 119).

Following Ophuls' view, which draws on Hobbes' theory, people have to choose between Leviathan and Oblivion. "If scarcity is not dead, if it is in fact with us in a seemingly much more intense form, ever before in human history, how we can avoid reaching the conclusion that Leviathan theory is inevitable? Given current levels of population and technology, I do not believe that we can. (...) Otherwise, the collective selfishness and irresponsibility produced by the tragedy of the commons will destroy the spaceship, and any sacrifice of freedom by the crew members is clearly the lesser of evils" (Ophuls, 1973, p. 224).

Ophuls' voice was not the only one; other scholars of the early wave of eco-political theory in the 1970s have used similar words, and have stated that they were "disgruntled with liberal democracy". Therefore, some of these theorists, such as Hardin and Heilbroner, have asserted that the protection of the environment and long term human survival require eco-authoritarian solutions.[38]

The point of departure with regard to the authoritarian perspective implies trust in the coercive powers of the State in a way that involves the reduction, if not total abolition, of traditional constitutional boundaries.[39]

The guiding idea was that the ecological crisis requires an "extraordinary concentration of power capable of suppressing human wants that, if left unchecked, would overwhelm the carrying capacity of the earth" (Torgerson, 2008, p. 18). Thus, the argument is that democratic government is not enough; a concentration of power is able both to overtake opposition from a disobedient and non-environmentally friendly citizenry, and to establish a policy for the common good.

---

38 For Heilbroner this approach demanded right, centralized government environmental regulation, energy and resource rationing, population control and a suspension of normal channels of political participation where these were seen to interfere with a swift and decisive governmental responses to the ecological crisis. See Heilbroner, 1974.

39 The works of Ophuls, Harden, & Helibroner are associated with the authoritarian perspective. See Walker, 1988, p. 67.

Moreover, Hardin, in his famous essay *Tragedy of the Commons*, went on theorising the most explicit conceptualisation of illiberal authoritarian politics (Hardin, 1968, p. 1243; Hartmann, 1992, p. 49). He warned about the impending perils of unrestrained and uncontrolled natural resource exploitation and environmental mismanagement: individuals are predisposed to over-exploit resources and pay no attention to the damage their economic actions cause on the environment. Gleditsch and Sverdlop (Gleditsch, & Sverdlop, 2003, p. 70) note that Hardin does not encourage confidence in the effects of economic and political freedom on environmental quality, because as he states, "freedom in the commons brings ruin to all" (Hardin, 1968, p. 1244; Hartmann, 1992, p. 49).

This latter argument is present also in Hobbes' Leviathan. According to Hobbes, the unrestricted freedom of humans in the state of nature leads to an inherently unstable and dangerous situation. Because of the individuals' striving for power and freedom, the collective good of preservation of life and the security of existence cannot emerge. In this situation there will be "continual fear, and danger of violent death; and the life of man, solitary, poor, nasty, brutish, and short" (Hobbes, 1974, p. 186).

The only solution is the establishment of an encompassing central power; the State or Leviathan can provide a way out of this *impasse*. Hardin has applied this kind of reasoning to the use that is made of the "commons space" in our world, where individuals are likely to show "parasitic behaviour with respect to the common spaces on the earth, since they are prone to reason with their own interests in mind" (De Geus, 1996, p. 190). The egoistic actions of the participants will, according to Hardin, "inevitably produce an environmental tragedy, unless people are prepared to consent to a system in which societal responsible behaviour can be enforced" (De Geus, 1996, p. 190). Hardin underlines also that the ecological crisis is also relative to the growth of inhabitants of the planet.[40] However, a solution to this problem cannot be found through purely scientific means, but instead looking at politics for answers.[41] Hence, to avoid

---

40 "The most important aspect of necessity that we must now recognise, is the necessity of abandoning the commons in breeding. No technical solution can rescue us from the misery of overpopulation. Freedom to breed will bring ruin to all. at the moment to avoid hard decisions many of us are tempted to propagandise for conscience and responsible parenthood. The temptation must be resisted, because an appeal to independently action consciences selects for the disappearance of all conscience in the long run, and an increase in anxiety in the short. The only way we can preserve and nurture other and more precious freedoms is by relinquishing the freedom to breed, and that very soon. Freedom is the recognition of necessity and it is the role of education to reveal to all the necessity of abandoning the freedom to breed. Can we put an end to this aspect of the tragedy of the commons". See Hardin, 1968, p. 1243.

41 Anti-democratic tendencies have also emerged in the activism of green groups such as Earth First!, with published an article citing the benefits of the AIDS epidemic to the

the inevitable tragedy, the freedom of private individuals must be drastically restricted and many of the rights we take for granted must be eliminated.[42]

The consequence of this model is the rejection of the idea of liberal democracy, in which the citizens are viewed as individuals acting to accomplish their objectives in a framework regulated by constitutionally granted legal rights and duties. For this reason, the model searches for possibilities to have more direct control over individuals.

To sum up the eco-authoritarian view, not only is the State a necessary condition to solve the current environmental problems, but even a State "along absolutist lines" is essential.

Nevertheless, even if the aim which this approach seeks to achieve is laudable, this proposal to sacrifice individual liberties is clearly not acceptable for several reasons.[43] Firstly, Ophuls' starting point is that citizens are the main polluters in society, and not private companies and State enterprises which are instead responsible for the overwhelming part of environmental pollution, rather than ordinary consumers themselves. Moreover, the scholar states that there is no substantial difference between an absolute State and a powerful and actively intervening State. A State could just take stringent measures in the area of the environment, without transforming itself in a Leviathan.

Finally, this view does not take into account that when the State becomes super powerful, it risks turning into a dictatorship and as a consequence, sacrificing individual liberties under the aegis of environmental protection.[44] Moreover, a State which restricts the citizens' rights in such a manner undermines the citizens' control over the non- environmentally friendly measures taken by the State itself.

2.1.2 Anarchical Perspective

The anarchist answer to the environmental emergency was not the envisioned conservation of a hierarchy as proposed by the authoritarian approach, but instead this hierarchy has been considered "the problem, not the solution"

---

ecology, suggesting that the disease should be allowed to run its course. See Dobson, 1995, p. 62.

42 Also Heilbroner argues that population growth poses a primary threat to environmental quality. Autocracies can curtail human reproduction, but democracies respect citizen rights, including that of procreation. See Heilbroner, 1974.

43 This environmental thought was defined "Hobbesian deeply despairing and anti democratic" by Hay. See Hay, 2002, p. 174; Heilbroner, 1974; Ophuls, 1977.

44 For instance the State could introduce the regulation which could diminish drastically the human rights and freedom.

(Torgerson, 2008, p. 18). Indeed, the anarchist perspective was based on the idea that the true origins of ecological problems were not uncontrolled human desires, as supported by the authoritarian model, but "hierarchical social structures, able of distorting the human potential to create cooperative communities in ecological harmony with nature" (Kenny, 1996, p. 23).

According to the anarchist approach, the crisis requires an "institutional transformation toward a pattern of decentralised, egalitarian and self-managing local communities attuned to ecological constraints and complexities" (Torgerson, 2008, p. 18). Hence, decentralisation became a fundamental concept: Bookchin (Bookchin, 1988, p. 96-97),[45] the most important eco-anarchist, gives three arguments for a fully-fledged decentralisation of society.

First, according to Bookchin, the restoration of human–nature relations requires creation of a society in which every individual is capable of participating directly in the formulation of social policy; and that this must be preceded by a removal of social hierarchical structures and domination. This requires a return to or creation of a society composed by relatively small autonomous units (Wilkinson, 2002, p. 216).

Secondly, decentralised communities can be sensitively adapted to natural ecosystems. Small decentralised communities do not destroy the natural landscape and can live in harmony with their surroundings.

Finally, there are the logistical advantages of decentralisation, for instance energy supply and transport. A big city uses immense amounts of resources in the form of oil, gas and coal. In his opinion the energy from sun, wind and tide, in most cases "can be provided only in relatively small quantities and for that reason are less suited for large cities" (De Geus, 1996, p. 194).

Hence, a fundamental point in this perspective is the abolition of the State. Such a transformation would be a revolutionary project, supported by a multiplicity of rising social movements, of which the green movement would be of crucial importance[46] (Torgerson, 2008, p. 18).

This proposal includes good suggestions which can be used for their own merits, nevertheless, this is not a 'panacea' for the ecological crisis for the

---

45 Bookchin: "If homes and factories are heavily concentrated, devices for using clean sources of energy will probably remain mere playthings; but if urban communities are reduced in size and widely dispersed over the land, there is no reason why these devices cannot be combined to provide us with all the amenities of an industrialized civilization. To use solar, wind and tidal power effectively, the megalopolis must be decentralized. A new type of community, carefully tailored to the characteristics and resources of a region, must replace the sprawling urban belts that are emerging today". See also Bookchin, 1990; M. Taylor, 1982.

46 The classic statement belongs to M. Bookchin: "Ecology and Revolutionary Thought", 1971.

following reasons: according to Wilkinson, first, "it is not correct to identify hierarchical relations as the basis of the overall environmental problems, rather, one could suggest that reformation of social relations should go hand-in-hand with or perhaps even follow resolution of human nature relations". Secondly, anarchist solutions do not necessarily provide any good grounds for the resolution of either because "feudal and tribal societies both typically small scale – can be exploitative of both people and nature, or networks of small communities may not provide an effective mechanism to prevent environmental harm" (Wilkinson, 2002, p. 216).

Another remark is that the local communities work really well on specific problems but lack the general overview of the total ecological situation and will probably also miss out on complicated and expensive expertise in environmental matters which can be generated by large centralised organisations. The risk could still be of societal and ecologically irresponsive behaviour; indeed, as Goodin affirms "a decentralised social organisation is more responsive to 'feedback mechanisms' and the people on the ground know more about the situation than those in control". On the other hand there is "the problem of local communities attempting to free-ride and avoid ecological policies. When there is insufficient mutual adaptation and a lack of centrally guided co-ordination of environmental polices there is a danger that the vulnerable ecological equilibrium of society might be violated".[47]

Another disadvantage underlined by this current thought is that there is a very short distance between those governing and those being governed and this situation could likely entail that those who govern might be reluctant to take disagreeable measures and, for example, instead try to pass on the locally produced pollution to the surrounding areas (Eckersley, 1992, p. 173-174).[48]

This point could be avoided if the anarchist proposal constituted merely an aspect of the new form of legal system in which environmental goals are better achieved by taking decisions at different levels. This system has to encompass the rights and the duties of the citizens. As it will be further explained, the recognition of citizens' environmental rights could avoid the problem. For instance, if a government at a local level takes environmentally unfriendly

---

47 See Goodin, 1992, p. 166, as it has been quoted by De Geus, 1996, p. 196.

48 Eckersley: "Historically most progressive social and environmental legislative changes [...] have tended to emanate from more cosmopolitan central governments rather than provincial of local decision making bodies. In many instances, such reforms have been carried through by central governments in the face of opposition from the local community or regional affected – a situation that has been the hallmark of many environmental battles".

measures, the local citizens or organisations would have the right and the duty to take actions against it.

As a consequence, the anarchist proposal to abolish the State would only partially resolve the ecological crisis.

Nevertheless, some interesting points proposed in this solution can be part of a multi-layered approach in which decisions are adopted at the level most appropriate to their ecological effects.

2.1.3 Democratic Perspective

Is it possible to save and protect the environment without installing an "eco-dictatorship" as William Ophuls suggests, or an eco-anarchy as Bookchin proposes? Despite disagreement among scholars,[49] the most accepted path is to maintain a democratic perspective while correcting its insufficiencies. A lack of democracy is the root of numerous ecological problems (Rocheleau, 1999, p. 38).

There are many reasons to improve this model; for instance Schultz and Crockett (Schultz, & Crockett, 1990, p. 53) and Payne (Payne, 1995, p. 41) remark that democratic rights and specifically freedom of information help to promote environmental groups, raise public awareness and encourage environmental legislation. Democracies are more responsive than autocracies to the environmental needs of the public (Kotov, & Nikitina, 1995, p. 17), and fulfil their obligations as contained in international environmental treaties and agreements (Neumayer, 2002, p. 139).[50]

---

49   In fact, the consequences of democracy on the environment are debated. Some scholars argue that democracy improves environmental quality, while others deduce that democracy increases environmental degradation or leads to environmental policy inaction where an environmental crisis is concerned. As Plumwood observes "it is matter of widespread observation that actually- existing liberal-democratic political systems are not responding in more than superficial ways to a state of ecological crisis which everyday grows more sever but which every day is perceived more as normality"; See Plumwood, 1999, p. 185. Moreover, has been maintained by Midlarksy that democracy is often connected with policy inaction where environmental devastation is concerned. He hold that first, there is the propensity of democracies to please competing interest groups. "Corporation and environmental groups can fight each other to a standstill, leaving a decision making vacuum instead of a direct impact of democracy on the environment. As the result of budget constraints, democracies may not be responsive to environmental imperatives but to more pressing issues of the economic subsistence of major portions of the voting public". Second, democracies may be indisposed to improve the status of environment, since some interest groups, as for instance industrial associations, are expected to have a better standing through democracy than environmental policies. See Midlarsky, 1998, p. 351.

50   See also Weiss, & Jacobsen, 1999, p. 16; Berge, 1994, p. 187.

Additionally, according to Gleditsch and Sverdlop, democracies respect human life more than autocracies and this is pragmatically necessary for states to become more ecologically sound.

Another advantage, as has been remarked by Lietzmann and Vest, is that to the extent that democracies engage in fewer wars, they should also have a higher level of environmental quality, because generally war devastates the environment (Lietzmann, & Gary Vest, 1999; Gleditsch, 1998, p. 381). Finally, Sen points out that famine tends to support environmental degradation for it diverts public attention from longer-term environmental concerns (Sen, 1994, p. 31). Since famines typically do not occur in democracies, environmental quality is expected to be higher in democracies than in autocracies.

Thus, this form of government is actually the only solution to the ecological crisis,[51] not just theoretically but also empirically, and even though there are studies on the outcome of democracy indicating that if a rise in the level of democracy leads to economic growth, democracy could incidentally cause more environmental degradation at the initial stage of development, it will help reduce it later (Li, & Reuveny, 2003, p. 29).[52]

According to a recent World Bank study, democratic countries tend to show a greater commitment than non-democratic countries to environmental policies. The aspect of democracy closely related to environmental policies is participation (Gates, Gleditsch, & Neumayer, 2003). At the political level, more and more States have acknowledged the existence of the link between democracy and environmental protection and they enshrined it in international law signing the Aarhus Convention which will be discussed later.[53]

If democracy is a non-negotiable element for a state to be more responsive to the environmental crisis, then how might this form secure the environmental goals? As said above, although deliberation and participation are distinct elements, the radical 'democratisation of democracy' can only succeed if participation and deliberation are regarded as two key elements in the process of collective decision-making. From the environmental viewpoint, these forms of democracy will better achieve environmental goals and build up a new form of democracy: environmental democracy.

---

51   Stein, 1998, p. 420. Beckerman, 1999, p. 85, (arguing that countries with the worst environmental records are the ones with the least respect for basic liberties and human rights); Torras, & Boyce, 1998, p. 147 (arguing that political rights and civil liberties have strong effect on environmental quality).

52   See also: Krutilla, & Reuveny, 2002, p. 23; Panayoto, 2000; Reuveny, 2002, p. 83.

53   The UNECE Convention on Access to Information, Public Participation in Decision-making and Access to Justice in environmental Matters, usually known as the Aarhus Convention, was signed on June 25, 1998 in the Danish city of Aarhus. Pallemaerts, 2003a.

2.2 Environmental Democracy: Formal and Spatial Dimensions

2.2.1 The Form of Environmental Democracy

The central issue is not whether the democratic State shall be abolished and replaced by another form of governance, but how this model can be adapted in such a way that more effective environmental policies can be carried out: how can the democratic State be changed into an environmentally protective and radically democratic model? How might it begin to reform and restructure actual political institutions so that they are more sensitive to environmental considerations?

To answer these questions, the starting point has been to observe the failure of representative democracy, and "the activities, backgrounds and interests of political representatives and decision makers are seen as far removed from the lives and perspectives of citizens. Although periodic elections act the mandate that representatives enjoy extends over a period during which time citizens have very little impact on decisions made in their name" (Graham, 2003, p. 53).

In deliberation and decisions, representative democracy inadequately bears "the plurality of environmental values as well as the interest of non-nationals, future generations and non-human nature" (Graham, 2003, p. 53), and it systematically under-represents ecological preoccupations. On the one hand, it represents only the citizens of territorially bounded political communities (Eckersley, 1996, p. 214), and on the other hand, the interest in environmental protection is systematically traded-off against the more immediate demands of capital and labour. Moreover, citizenship, as it has been understood in representative democracy, is usually a passive affair which has led the electorate to a "spread of cynical attitudes about public affairs and the notion of a public good" (Offe, & Preuss, 1991, p. 165).

Theories of participatory and deliberative democracy also offer an interesting theoretical response to environmental problems.[54] Both provide institutions

---

[54] See in general: Arrhenius, 2007; B. Barry, 1978, p. 204; J. Barry, 1999; Beckman, 2006, p. 153; S. Chambers, 2003, p. 307; Dobson, 1996, p. 125; Dryzek, 2000; Eckersley, 2004; Farrelly, 2004; Fish, 1999, p. 88; Goodin, 2003; Goodin, 2007, p. 40; Gutmann, & Thompson, 1996; M. Jacobs, 1997, p. 211; G.F. Johnson, p. 67; Kumar, 2003b, p. 99; Linklatera, 1998; Macedo, 1999, p. 3; Nagel, 1986; Page, 2006; Parfit, 1987; Rawls, 1971; Rawls, 1993; Reiman, 2007, p. 69; Scanlon, 1998; Sharander-Frechette, 2002; G. Smith, 2003; D. Thompson, 2005, p. 245; Tully, 2002, p. 204; Whelan, 1983, p. 13; Weinburg, 2009; M. Williams, 2000, p. 124; Woodward, 1986, p. 804; Young, 1990; Young, 1999, p. 151; Young, 2000; O'Neill, 2002, p. 257.

that support democratic deliberation, which will be aware of the "plurality of environmental value and which promote political judgments that takes into consideration different perspectives on the non-human world and promises a political environment within which the plurality of environmental values can be effectively and sensitively assessed and considered in decision-making processes" (Graham, 2003, p. 54). Decision-making should be arrived at by processes of communication between all people affected by an issue rather than by coercion, deception or insensitive representation" (Kitchen, Milbourne, Marsden, & Bishop, 2002, p. 139).

Both believe that citizens may "contribute intelligently and reasonably to politics, especially when they know their contributions matter, and most believe that citizenship should be able to contribute to more areas of decision-making than they can now" (Hauptmann, 2001, p. 399; Baber, & Bartlett, 2005, p. 255). Participation, indeed, has been seen as vital to the flourishing of democracy (Pateman, 1970, p. 43). At the origin of this approach there is Kelsen's view: "political freedom, that is, freedom under social order, is self-determination of the individual by participation in the creation of social order" (Kelsen, 1961, p. 285). It is certainly "the case that an uninformed and uninvolved community cannot adequately protect its environment and natural resources" (Douglas-Scott, 1996, p. 113). Indeed, participation offers people the opportunity to make decisions concerning their own environment.

Formally, direct control might be established by making the outcome of an appropriately constituted participation process legally binding, by requiring that there be a presumption in favour of the outcome chosen by participants, or by allowing the public to directly veto some options. Moreover, the participation in environmental decision making, should also seek to give some direct control to the public to avoid "secrecy" (M. Jacobs, 1999, p. 114). Jacobs affirms that: "What creates anxiety and feeds mistrust is secrecy, the sense that experts and politicians are making decisions without public scrutiny, and possibly subject to bias from business lobbying interests, these writers advocate new techniques of public consultation and debate, such as citizens' juries, consensus conferences and deliberative polling, and argue that politicians must become much more sensitive to underlying public anxieties if they are to avoid disasters of governance of the kind witnessed in recent years".

Also, the participation through public consultation is an essential feature of all environmental decision making processes. Both control and consultation reflect the ideal of democracy and may promote sustainability under the right conditions (Bell, 2004, p. 106). It suggests conditions under which citizens can meet and exchange views on environmental knowledge and value, and it is more possible to incorporate these in their judgments and practices.[55]

Moreover, deliberative processes present a favourable arena in which citizens can find alternative ways of conceptualising relations between human and non-human worlds and ethical issues are discussed in the public domain (Heyward, 2008, p. 625).

Indeed, deliberative democracy is useful to build an environmental democracy because it seeks to educate through "dialogue and transform political opinion through reasoned debate, rather than simply aggregate the sum of unchallenged individual desires" (Heyward, 2008, p. 625). It is worth noting that to protect the environment, the educative effect of participation as consultation "will only influence societies in the long term" (Bell, 2004, p. 108). Consequently, the democratic model is more demanding than liberal democracy because it requires more time, patience and information (Heyward, 2008, p. 625).

Because of the aforementioned characteristics, the potential merits of deliberative democracy have been criticised, for this model, in most aspects, continues to be extremely abstract and theoretical. Nevertheless, deliberative democracy has to be understood as a constructive project, and consequently it is necessary to institutionalise it by law. Graham has suggested three potential deliberative instruments that could encourage augmented participation and deliberation by citizens in the decision-making process and help to enhance

---

55   As John Barry notes, there is a good deal of support in the literature for the view that the form of democracy which best fits with green politics is a deliberative model: J. Barry, 1999.

practices in verbalisation, enforcement and evaluation of environmental policy: citizens' forums, [56,]referenda and other citizens' initiatives.[57]

---

56   There are three types of citizens forums namely deliberative opinion pools, citizens' juries and consensus conferences, which provide the space for citizens to deliberate on pressing policy issues. Those types share a number of features: across-section of the population is brought together for three to for days to discuss an issue of public concern; citizens are exposed to a variety of information and here a wide range o views form witnesses whom they are able to cross-examine; and fairness of the proceedings is entrusted to an independent facilitation organisation. For example a trial citizen juries on waste policy in Ireland, Flynn, 2008, p. 57. Flynn discusses the promises and the limits of planning cells and citizens juries for environmental decision-making, illustrated with comparative evidence and an Irish case study concerning waste policy. See also Kenyon, Nevin, & Hanley, 2003, p. 222: "A citizens' jury consists of a small group of people, selected to represent the general public rather than any particular interest-group or sector, which meets to deliberate upon a policy question. Although relatively new in the UK, CJs developed independently in Germany and the US in the early 1970s to advice on a range of issues including planning, health care and political issues. In both cases, CJs were considered a tool which could be used to enhance democratic and administrative processes. The success of CJs in Germany and the US has seen CJs and other small-group decision-making approaches gain in popularity in many other Western countries, including the UK and Australia [...] Initially, CJs were concerned with health care issues; however, more recently, they have been used to address environmental issues. In the US, for example, a CJ was asked to rank environmental risk and in the UK a number of CJs concerned with environmental issues have taken place, including one in Hertfordshire on waste management and one in Ely on the creation of wetlands". "The key advantage that citizens' juries have over other environmental decision support tools is that they provide policy-makers with input from a group of well informed and representative members of the community. Jurors become well informed because information provision, time, scrutiny and deliberation are all crucial aspects of the process. A second advantage that the uses of CJs has over other methods of public opinion gathering is that it allows consumers to be asked what Sagoff and Jacobs might call 'the right question'. It asks them to deliberate on the environmental issue in terms of what is best for society. Thirdly, the notion of value construction suggests that respondents do not have well defined preferences for many complex policy options, but that these preferences are constructed during the elicitation process itself. The way in which data about environmental preferences are collected is therefore very important. Gregory suggest that approaches which encourage participants to construct their preferences and reveal their thinking as part of the information-gathering process, yield more detailed information about key attitudes and trade-offs, which can then inform the decision-making process. A CJ approach will be particularly useful in this as it consists not just of presenting information to participants, but also asks participants to carry out a variety of tasks throughout the process. Using tasks within the citizens' jury process helps the participants to construct their preferences in a rational and transparent manner". At p. 230: "As Crosby outlined, the role that citizens' juries play in environmental decision-making is to provide policy-makers with input from a group of well informed and representative members of the community with which to advise the decision-making process. CJs therefore offer one way in which decision-makers can obtain quality information about public views on important issues, and a means of devising innovative solutions".

57   Both scholars' method are two processes by which a population can vote directly on policy issues. Referendums can be advisory or mandatory and the initiative offers a process through which citizens are able to put forward new legislation or nullify existing

The underlying idea is that an informed and legally empowered citizen is the most important aspect of environmental democratisation. In order to achieve it, deliberative democracy has to be integrated with participative democracy which acknowledges a range of procedural rights, ranging from the right to information, to the right of legal redress and the rights of participation.

As regards these rights, Graham asserts that "democratic deliberation cannot effectively progress without adequate environmental information, much of which is held by public and private authorities. Legal redress offers the opportunity to object to decisions and actions of public and private bodies that contravene environmental rights and law". Environmental law acts as a "legitimate constraint on the outcomes of democratic policy making" (Graham, 2003, p. 108).

According to Eckersley, "the introduction of environmental rights clearly has the potential to alter radically the established framework of decision making in favour of the environment".[58] Such procedural rights "would not only help to redress the current under-representation of environmental interests but would also provide a firmer guarantee of environmental decision making according to law –thereby redressing the pervasive 'implementation deficit' in environmental law and administration".[59]

In conclusion, it can be said that to develop environmental democracy the most advantageous form is the democratic one, which in particular includes deliberative and participatory tools.

However, environmental democracy has not only a formal dimension but also a spatial dimension as will be explained in the following paragraphs.

---

laws. See also G. Smith, & Wales, 2000, p. 51; Renn, 1995.

58 Eckersley, 1996, p. 216, The author proposes to concretise this right through the introduction of an environmental bill of rights embodied in ordinary legislation or the constitution which declares that citizens have a right to ensure that environmental quality is maintained in accordance with the standards set by current environmental law. See also Hayward, 2000, p. 563.

59 Eckersley, 2000, p. 230. Again Eckersley has also suggested a different kind of procedural constitutional mechanism to ensure the enforcement of substantive environmental rights: the constitutional entrenchment of the precautionary principle provided by the principle 15 of the Rio Declaration which states: "Where there are threats of secures or irreversible damage, lack of full scientific certainty should not be used as a reason for postponing cost-effective measure to present environmental degradation". As Graham has suggested the "principle would act as a procedural norm in the policy-making process and could also benefit citizens seeking legal redress against decisions that generate potential environmental risk, because the burden of proof would be on the defendant to show why preventative action is not necessary". Graham, 2003, p. 111.

2.2.2 The Space of Environmental Democracy

2.2.2.1 Global Level

The second dimension of the environmental democracy is the *spatial one*. Environmental problems can be divided into two categories: those which are principally addressed by global management; and those which are primarily addressed by local management.

The first group of problems consists of the pollution of the global commons and its consequences. Indeed, recently, advances in scientific knowledge have revealed forms of pollution that have an impact far beyond the boundaries of border regions, international catchment areas and regional seas.

These global ecological problems are not confined to a limited number of states, but affect the international community as a whole, for both their causes and their effects are widespread and indirect, defying all political and legal borders (Pallemaerts, 2003a). These commons, enjoyed by all, include elements of the biosphere and natural resources (Pallemaerts, 2003a) e.g. the oceans, the atmosphere (Robinson, 1972, p. 44) and Antarctica, the Polar Regions, and outer space.[60]

Moreover, the globalisation of environmental issues does not simply reflect the fact that there is only one biosphere, but also has a consequence in economic globalisation. Unsustainable production and consumption patterns, which are the cause of most ecological problems, are increasingly shaped by economic processes and players that transcend territorial boundaries and are therefore outside the control of individual states. As a result, there are "very few environmental issues, these days, that cannot be described as international, in terms of either their causes or effects" (Pallemaerts, 2003a).

The solutions, in order to manage such problems, have to be sought at the global level, through a global environmental democracy, or as it has been called

---

60   There is agreement about that climate change constitutes a global problem because "it is caused by, affects, and cannot be remediated without the participation of a multitude of global actors. The global nature of the causes of the issue also implies that national boundaries, the traditional realm of citizenship operation, are merely another obstacle to effective action on the problem, as demonstrated by the lack of international cooperation on the ratification of the Kyoto Protocol by the world's most significant emitter, the United States" and the failure of the Copenhagen Conference in December 2009. The impacts of Climate Change are unequally distributed across geographic space and in time, and they do not respect national boundaries, impacting those that are most vulnerable. Wolf, 2007; Agarwal, & Narain, 1992, p. 7. See also recently: Balck, 2010, p. 359; Karassin, 2010, p. 383; Takacs, 2010, p. 521.

by Eckersley through a "transnationalised environmental democracy", which responds in particular to transnational problems (Eckersley, 2004, p. 197-198).

The most realistic means of extending the global dimension of democracy is through multilateral agreements between States, which generate overlapping, additional rules-systems, which on their behalf dynamically employ existing territorial governance rules and structures. Over the last fifteen years, concerted international action has begun to address problems on a global scale.

There are already several multilateral initiatives that have moved along this path, which not only prove the realistic possibility of such an approach but also the realistic possibility of emergence of transnational states. The most significant example is the Aarhus Convention, which, as the UN Secretary General has described it, is the most "ambitious venture in the area of environmental democracy" (Annan, 2000; Bell, 2004b, p. 94).

In this sense, it represents the first international convention dedicated to creating trans-boundary environmental rights of individuals in the move towards an environmental democracy (Eckersley, 2004, p. 193-194). Indeed, the individual dimension and its role in solving the global environmental issues are fundamental, as will be discussed in detail in Section II.

2.2.2.2 Local Level

Global environmental issues and solutions "require also local approaches" (Myint, 2003, p. 287).

The local level encompasses regional and national orders. At the level of each state, there are also problems which are really linked to that particular country. Moreover, many environmental problems extend beyond a single nation but are limited in their effects to one region. Such environmental problems have a localised impact on a region divided between different national jurisdictions; where rivers or lakes border different countries.

Thus, on a local level a single state or a regional organisation composed of different states must construct an environmental democracy to solve the second category of environmental problems, e.g. soil erosion, desertification, deforestation, water pollution, air pollution, and protection of nature parks, sanctuaries and areas of biodiversity. Indeed, local communities can be more suitable in determining and managing systems of environmental protection and sustainable use which function for the amelioration of such problems (Agarwal, & Narain, 1992, p. 1).

This last view encompasses some elements of the anarchical model of State; however, this does not mean that these problems can be left completely to local communities, as the approach suggests. Global vision has to play a decisive role and must help in making community management effective. Regional organisations and states become "local agents of the common good" facilitating trans-boundary democracy.

Hence, local environmental democracy refers to a community's right to manage its immediate environment through deliberative and participatory institutions. Moreover, every state should commit itself to build up a new system of governance within its borders. A system of local level governance, through open and participatory institutions, with inalienable rights to care for and use, would manage its immediate environment.

The promotion of environmental democracy by States or regional organisations should require that all the governments grant to their citizens a clear environmental right to a clean and healthy environment as well as procedural environmental rights. It is crucial that every citizen in the world should have the right to challenge in court any decision that affects her or his immediate environment.

2.2.2.2.1 National Local Level

As seen above, some local problems can be linked to a specific national territory which can concern the territory of one country or part of it, for instance concerning a city or the surrounding countryside. Cities are the principal perpetrators and the main victims of environmental damage. Indeed, "threats to the environment come mainly from urban emissions, motor vehicles and framing and process industries, especially chemical industries; urban water, soil and air are threatened, noise levels increase, and anaesthetic pollution rises" (Mega, 1997, p. 52).

To solve the environmental problems at this level some authors have noted that the linkages between city and country are fundamental: rural sustainability and urban sustainability are "two faces of the same coin", the achievement of a balanced, diversified economy and healthy ecology "by means of regional and rural development programmers, and the restructuring of fragmented natural urban areas which are threatened by urban development and re-urbanisation, are essential for the sustainable development of rural areas". The achievement of sustainability on a global scale requires the achievement of both urban and rural sustainability, which do not follow parallel ways, but interconnected objectives" (Mega, 1997, p. 47).

The environmental performance of a city is decisive for urban sustainability. It is closely linked to environmental management and depends on the participation of the local population. Thus, if an environmental democracy is to be established and to move towards this transformation, the recognition of environmental rights and duties is fundamental.[61]

Conclusion of Section I

To understand the form and space of the environmental democracy, it was first necessary to study the meaning of the words 'democracy' and 'environment', and give a brief history of the green debates concerning possible alternative forms of state (Eckersley, 1996, p. 214). With regard to the form, there is still a need to maintain a democratic model, but it is also necessary to modify the existing liberal democratic institutions (Eckersley, 1996, p. 213), however, not by such a radical change as authoritarian or anarchical views suggest.

Nevertheless, from a formal viewpoint, a gradual reconstruction of democracy will have to take place, using some valuable elements from the ecological visions of authoritarian and anarchist elements like command and control from the first model and decentralisation from the second. Concerning the spatial dimension, it is necessary to build up the new democracy both at the local and at a global level, in order to achieve environmental protection.

Environmental democracy must begin with a gradual transformation of the form, style and content of democracy, but also of society's relationship with the rest of nature.

Indeed, setting up systems of global and local management involves the promotion of a new concept of citizenship, in order to manage the problems which stem from the ecological crisis: environmental and ecological citizenship. In any democratic framework, citizenship is built upon a system of both rights and duties.

This new citizenship derives from the idea that there are specific rights and duties *vis-à-vis* citizens and in the case of the environment also rights and duties *vis-à-vis* future generations and nature itself. Active citizenship can be a source for the protection of the environment; it is even one of the essential conditions.[62]

In fact, if the individual is an active citizen, s/he feels independent on the one hand, but at the same time, s/he depends on her/his surroundings; s/

---

61 "Local agenda 21 and local governments. Two-thirds of the actions proposed by the Rio Conference and agenda 21 require the active involvement of local authorities [...] the Charter of European cities and towns towards sustainability, issued in Aalborg, states that sustainability development helps cities and towns to base standards of living on the carrying capacity of nature, while seeking to achieve social justice, sustainable economies and environmental sustainability ". Mega, 1997, p. 58.

62 Alcaeus suggests that "it is not the roofs, the stones of the wall, the canals that make a city, the city is made by men, able to enhance opportunities". Ceiner, 1984.

he experiences her/his liberty but then s/he becomes aware of his limits. S/he is therefore inclined to feel responsible for her/his environment and living conditions, and is urged to understand her/himself as a participant in the ecosystem, planet, region, state and city (Hall, 1997, p. 24).

Consequently, the next section will analyse the concept of citizenship in environmental democracy and will take up the challenge of one particular liberal democratic institution, namely the rights of the individual as well as a new institution, namely duties of the individual. The choice of the rights and duties has been used as a tool to connect the democratic form and the ecological issues.

The following approach will therefore be to identify seemingly feasible adjustments to existing state structures and mechanisms, and to show that "greening" the citizenry can improve the protection of the environment (Saward, 1998, p. 345)

Section II: The Actors of Environmental Democracy: the Environmental and Ecological Citizen

"Democracy requires that the government be not only of the people and for the people, but also by the people".
(Abraham Lincoln in Gemmill, & Bamidele-Izu, 2002, p. 77)

"No magic wand can be waved over the multitudinous problems of environmental quality
No elegant declaration of rights will simply and quickly solve our problems, protracted struggle lies ahead, and citizens fighting to vindicate their rights will be a central figure in that struggle. [...]
Effectuation of the public interest must begin to embrace the active participation of the public"
(Sax, 1972, p. XVIII-XIX).

The growing ecological crisis is pushing the citizens of Earth to realise that the world is just one and that it has to be used and protected (Agarwal, & Narain, 1992, p. 3). This situation emphasises the growing disjuncture between moral and legal citizenship as defined by the nation-state and it pushes towards a more universalistic, inclusive codification of guarantees of citizens' rights and towards a definition of ecological responsibility (Christoff, 1996, p. 151).

States seem to slowly realise that they cannot create green societies on their own, but that they have to recognise a role for civil society in the process of achieving environmental objectives. It can be assumed that 'sustainability' is one of these objectives, broadly committed to by governments around the world. Moreover, citizens as well as governments have a role to play in carrying this obligation.[63]

---

63 For instance the preamble of the Aarhus Convention affirms: "The Parties to this Convention [...] Affirming the need to protect, preserve and improve the state of the environment and to ensure sustainable and environmentally sound development," and "Desiring to promote environmental education to further the understanding of the environment and sustainable development and to encourage widespread public awareness of, and participation in, decisions affecting the environment and sustainable development".Dobson, & Saiz, 1998.

Hence, achieving ecological aims requires a process of democratisation through participation, taking into account that economic transformations, scientific-technological progress and daily life changes will not be enough. Citizens' participation in the environmental decision-making process is marked as essential, and as it already has been seen, this is possible only through modifications within the democratic model. The environmental participatory rights approach as well as the ecological duties approach towards the promotion of an environmental and ecological citizenship must be read in this context.

In most green political proposals, environmental citizenship is seen as a tool to include the individuals in political participation. In particular, stress is placed on the rights of access to information, participation and access to justice, as well as on democratic models. Moreover, as mentioned above, obligations *vis-à-vis* the planet and future generations are also emphasised (Melo-Escrihuela, 2008, p. 113).

From this perspective, section II, which will be dedicated to the actors of the environmental democracy, has been divided into two parts. After an analysis of the theoretical construction of a new citizenship,[64] the first part, in particular, will deal with the environmental citizen and her/his corresponding substantive and procedural environmental rights. The second part will focus on the ecological duties corresponding to ecological citizenship. It has to be emphasised that notions of ecological duties or obligations, from a legal point of view, are more difficult to elaborate.

According to a legal analysis, individuals have the general duty to respect the rights of others and to abstain from disobeying those social customs codified in laws. Rights can subsequently be identified as the primary focus of attention since they stand logically prior to duties. Rights are also more tangible than duties since they benefit from a higher degree of public visibility, understanding and support than a parallel discourse of duties (Feinberg, 1988).

Of course, duties are easily conceptualised in ethical terms based upon arguments of political philosophy rather than upon legal terms. Nevertheless, the purpose of this book is to try to move on from this construction of duty towards a legal approach. In fact, the first step to the recognition of human rights was also taken through the establishment of a philosophical and ethical basis (Hancock, 2003, p. 3).

---

64   The first Author that spoke about environmental citizenship was M. J. Barker, 1970, p. 33-35.

In this regard, it is helpful to look at the notions of citizenship from a completely different perspective, turning to conceptions of citizenship based on moral responsibility and participation in the public sphere (Christoff, 1996, p. 151).

1 A Theoretical Conception of New Citizenship

As Environment Canada has put it:
"Each of us has an effect on the environment every day; the key is to make this
impact a positive one.
We must all take responsibility for our own actions,
whether as individuals, or as members of a community or an organisation.
Let's work together and become good environmental Citizens!
If you don't, who will?"
(Environment Canada, 2004).[65]

"Au-delà des différences de formulation, la présence fréquente du binôme droit-devoir montre bien qu'en matière d'environnement, 'chacun' est à la fois victime et pollueur, que la protection de l'environnement implique la responsabilité individuelle face à des préoccupations collectives. Le droit à l'environnement se situe ainsi à l'interface de l'intérêt général de protection de l'environnement et de l'intérêt subjectif à la sauvegarde de la qualité de l'environnement"
(Van Lang, 2007, p. 123).

Several theories have been developed regarding the different ways in which citizenship and environment might be related. On the one hand, sociologists have explored the issue of environmental identity; on the other hand, political theorists have discussed issues of membership in relation to environmental citizenship (Hilson, 2001, p. 336).

Concerning the latter, citizenship in an environmental democracy has been defined in different ways, each term involving different features: for instance "ecological citizenship" (Christoff, 2005, p. 481; Milton, & Curtin, 2002, p. 293), "green citizenship" (Hartley, 2001, p. 490), "environmental citizenship" (Dobson, & Bell, 200; Luque, 2005), "sustainability citizenship" (J. Barry, 2006; Hay, 2002), "environmentally reasonable citizenship" (Hailwood, 2005, p. 195) or "ecological stewardship" (J. Barry, 2002, p. 133). For the purpose of this book, however, it is more useful to examine and define the diverse, but complementary, concepts of environmental and ecological citizenship.

Hence, this analysis will start with elaborating on the interesting distinction between environmental and ecological citizenship made by Andrew Dobson (Dobson, 2004, p. 116), who is amongst the most representative theorists of this trend, especially since his publication *Citizenship and the Environment*.

---

65   Available at www.ns.ec.gc.ca/udo/who.html

To better understand this new form of citizenship, it is worth remembering that since Marshall[66] it is customary to distinguish between three forms of citizenship: civil, linked to the right to associate, to speak freely; political, linked to the right to vote and run for election; and social, linked to welfare rights. Moreover, two types of citizenship are differentiated: liberal and republican. Nowadays, however, three forms and two types of citizenship are not sufficient, since the environmental changes taking place in the world oblige us to rethink the spatial framework of citizenship.

Thus, Dobson has added "environmental form" to the list of forms and "post-cosmopolitan"[67] to the types of citizenship, which leads to two new aspects of this new citizenship: an environmental and an ecological dimension. Hence, a distinction needs to be drawn between them.

The former can almost entirely be expressed in the language of the two major traditions of citizenship, liberal and civic republican, but an environmental form has been added.

The latter does not consist of a liberal or republican type, but it is a 'post-cosmopolitan' citizenship that has four principal characteristics. First, it deals with a non-reciprocal responsibility; second, it works with a non-contiguous and non-state understanding of political space, best understood in terms of the 'ecological footprint'; third, it argues that the private arena is as much a sphere for citizenship as the public arena; and, finally, it links to the notion

---

66 For that see Burker, & Rees, 1996.

67 To know more about cosmopolitan citizenship see: MacGregor, 2004, p. 85. "There has been growing interest in cosmopolitanism since the early 1990s. In response to economic and cultural globalisation and the on-going existence of national, ethnic, and religious conflict, many academics and activists involved in social change movements are seeking ways to make thinking and acting globally an ethically compelling approach to politics. Cosmopolitanism, a Universalist ethical perspective with roots in ancient Stoic and Christian traditions, seems to provide precisely this kind of stance. The origin of the term cosmopolis is the link between cosmos, the order of the universe, and polis, the order of society. It is more recently associated with the 18th-century philosopher Immanuel Kant, who believed that all human beings are endowed with a unique capacity for moral behaviour and that because respect for humanity is part of universal natural law, 'morality should be supreme over politics'. Worried about war and other negative effects of nationalism in world politics, Kant argued that our sense of common humanity unites us in a more fundamental way than loyalty to rulers or states. Therefore, it is the duty of every human being to work toward a cosmopolitan society which would lead to perpetual peace. Those who are inspired by Kantian thinking today find in cosmopolitanism a framework for global ethics that speaks to individuals, as 'citizens of the world' with universal rights and responsibilities, and an overarching institutional–legal model for peaceful global politics Contemporary proponents of cosmopolitanism often look to the environmental crisis, which includes transboundary problems like ozone depletion and nuclear waste, as added justification for developing a universal ethical perspective, international law, and practices of 'global environmental citizenship".

of citizenship virtue, understood in terms of the virtues required to meet an ecological citizenship's specific obligations rather than in terms of 'citizenship virtues' as they are more arbitrarily understood.

Moreover, environmental citizenship corresponds mainly to a rights approach, also called rights-based, as well as a duty-oriented approach.

The rights-claiming approach acknowledges the multiple layers of rights that individuals, groups and communities enjoy. Bell argues that these rights enable individuals to make choices and exercise their power in their everyday lives in addressing environmental matters (Bell, 2005, p. 179).

These rights minimally include environmental goods such as clean air and water, but also include procedural rights in decision-making about the environment. In contrast, the duty-based aspects of ecological citizenship encourage people to take more responsible environmental actions and act differently for the sake of the environment (Carolan, 2006, p. 345; Wonga, & Sharpb, 2009, p. 37).

Duties include "the obligation to comply with environmental laws but also to justify choices about lifestyles which affect the environment" (Flynn, Bellaby, & Ricci, 2008, p. 769). Despite the different sphere of action of the two new approaches to environmental citizenship, they have been regarded as "complementary" for they can both plausibly be read as heading in the same direction: "the sustainable society" (Dobson, 2003, p. 82) which is at the basis of environmental democracy.

1.1 Environmental Rights

Dobson uses the term environmental citizenship to deal with the connection between citizenship and sustainability from a liberal point of view. Thus, environmental citizenship is conceived as an extension of liberal citizenship: "by enshrining human environmental rights and rights of participation in the constitutional context" (Melo-Escrihuela, 2008, p. 130). Environmental citizenship is mainly attached to the state's territory, and may be understood as an additional element of the traditional statement concerning the three elements of citizenship: civic, political, and social. Consequently, environmental citizenship implies a new set of human environmental rights.

Thus, environmental citizenship is articulated as a *status* that would be guaranteed by virtue of enshrining environmental substantive and procedural rights in laws (Jelin, 2000, p. 47). In this respect, the Aarhus Convention serves as an example of how a rights-based conception of environmental citizenship could be instantiated.[68] Hence, attempts to amplify citizens' participation in decisions regarding the environment, and projects to fortify the existing democratic institutions are part of what environmental citizenship should be about.[69]

This citizenship is linked to the dimension of environmental rights that is conducted exclusively in the public sphere, whose principal virtues are "the liberal ones of reasonableness and a willingness to accept the force of the better argument and procedural legitimacy, and whose remit is bounded by political configurations modelled on the national-state" (Dobson, 2001, p. 490).

The term environmental rights is often used as a common expression by many authors in the debate concerning the expansion of the rights-based approach to function as an environmental safeguard, encompassing both substantive and procedural rights.

Both types of rights have habitually been classified as beheld within the third group of human rights.[70] This category covers solidarity rights, and is also known as the third generation of human rights (Alston, 1982, p. 307). These rights are

---

68   The preamble of Aarhus Convention recognises that "Recognising that, in the field of the environment, improved access to information and public participation in decision-making enhance the quality and the implementation of decisions, contribute to public awareness of environmental issues, give the public the opportunity to express its concerns and enable public authorities to take due account of such concerns".

69   The preamble of Aarhus Convention recognises also that "in the field of the environment, improved access to information and public participation in decision-making enhance the quality and the implementation of decisions, contribute to public awareness of environmental issues, give the public the opportunity to express its concerns and enable public authorities to take due account of such concerns".

70   Human Rights: idea of norms which occupy a higher moral and legal status in a given hierarchy of norms due to their association with human dignity There are three broad groups: civil and political rights are those which provide individuals with rights and freedoms within their society (examples of these are the right to life, right to political participation); economic, social and cultural rights relate to the quality of life of individuals and communities (examples are the right to health and the right to education); and solidarity or third generation rights.
On the historicity of the protection of human rights see Bobbio, 1997. The following have also been included as likely third generation rights: The right to development. The right to peace. The right to ownership of the common heritage of humankind.
For Horn: "There continues to be doubt about the usefulness of the classification of the right to a healthy environment as a third generation human right. Pathak (1992, p. 205) suggests that there is no need to distinguish the human right to a healthy environment in this way as in fact it could be classified as a first, second or third generation human right. This is because this environmental right can form part of the human right to life Secondly, the right to a healthy environment can also be derived from the International Covenant on Economic, Social and Cultural Rights and so may form part of the second generation of rights. Thirdly, as these rights have an international dimension then they may also be located within the third generation of human rights. So the right to a healthy environment can overlap these three areas. There are areas which environmental law and human rights law have in common". Horn, 2004, p. 264.

usually correlated to groups of peoples, rather than to individuals. The aspiration is to grant to certain groups rights related to issues such as peace, development and environment.[71]

Therefore, it is fundamental to distinguish between these two categories of rights. While a substantive environmental right "would entitle the holder to a specific quality of environment, procedural environmental rights would entitle the holder to processes", such as: appropriate access to information concerning the environment; participation in decision-making processes; and access to justice related to environmental matters (Turner, 2009, p. 48).

1.1.1 A Substantive Environmental Right

1.1.1.1 The Reason to Recognise a Substantive Right

1.1.1.1.1 A Substantive Right Better Protects the Environment and Other Human Rights

It should be noted that a diversity of expressions has been employed to express a potential substantive environmental right. For example, a right to a "healthy environment", a "clean environment" and also a "right to environment" have been frequently expressed. The right to a "decent environment", a "safe environment", an "adequate environment", a "satisfactory environment", and a "viable environment" have been utilised as well. And the list is not complete, as there is a huge variety of other terms.[72] The common feature of all these expressions is that they have been used with regard to the potential development of a substantive environmental right.[73]

The employment of diverse expressions mirrors also the general absence of the accepted advancement of a universal substantive environmental right (Turner, 2009, p. 17). The problem concerns the question of definition because

---

71   Therefore they have the potential to fit into all three categories: for example, civil and political rights could be invoked to protect the right to life or procedural rights of participation. An example of economic and social and cultural rights being used may relate to an action brought to protect health standards owing to unwarranted levels of pollution. See Turner, 2009, p. 17; McCaffrey, & Lutz, 1978.

72   Recently Turner in his essay *A Substantive environmental Right*, has presented a draft, suggesting the formulation" the human Right to a Good Environment" that represents an attempt to encompass a right to a good clean and healthy environment in which all ecosystems and natural systems are protected for all peoples. Turner, 2009, p. 221-222.

73   They reflect in addition the fact that some authors examine the question in terms of an anthropocentric right while others include the environment, including ecosystem and natural systems.

the abstract and general formulations, such as the right to a clean environment, are in a similar category to abstract claims for a right to employment. Both claims are desirable, but it is not always easy to identify those who are responsible for causing the problem of pollution or unemployment. In addition, the absence of clear and settled standards of quality of environment provides the "Achilles' heel" (Eckersley, 1996, p. 228) in the case of environmental rights.

The definition-problem is not a convincing reason not to act more proactively, since a "certain extent of vagueness is common to each human right. Also it becomes not an impossible exercise when we identify its components more precisely" (Hectors, 2008, p. 174).

Consequent to this difficulty, several commentators have pointed out that the best way forward for environmental rights can be found in the provisions of procedural rights rather than of a substantive right to an adequate environment. Nevertheless, according to other authors, environmental rights should also be substantive and not merely procedural for numerous reasons.

First, the development of a limited autonomous right to environmental protection is necessary in order to be able to restrict overly wide national margins of discretion by national authorities in environmental matters (Hectors, 2008, p. 174); second, procedural rights are essentially participatory rights.

Unlike a substantive right to an adequate environment, they do not entail any direct obligation of the state regarding substantive environmental protection measures. Moreover, the fundamental difference is that the procedural rights involve a duty to refrain from an action, while substantive rights imply a duty to act. Procedural rights do not directly entail any substantive duty at all; in this regard, Hayward has affirmed that "it might be supposed that this would make procedural rights even less problematic to defend than negative rights" (Hayward, 2005, p. 84).

In fact, Saward suggests that a green democratic right could be expressed as a negative right, thus: "the State must not deprive citizens, or allow them to be deprived of the environment". Thus, just as individuals have the right not "to be subject to the kinds of harm wrought by practices of torture, unlawful deterioration, and so on, they may equally be thought to have a right not to be subject to comparable sorts of harm which might be wrought through practices which assail them, for instance, with toxic pollutants".

It therefore seems appropriate to view a right to an adequate environment as a negative right to the extent that the demand it implies is not that the government has to provide a clean environment, but that it prevents private parties- and their own agencies- from polluting or despoiling what would otherwise have been, without the need for any positive action, an adequate environment".[74]

1.1.1.1.2 A Substantive Right Better Protects "Environmental Refugees"

The necessity to recognise a right to environment has become important and also urgent, especially with regard to the issue of environmental refugees.[75] Even

---

74   "Nevertheless, according to Hayward there are several objections to this view. An initial objection "is that such a right might be advanced in circumstances where the environment has already been compromised, and so the demand based on it would not literally be preventative: the demand might be for rectification of or compensation for harm that has already been done rather than of prevention of some impending harm. [...] The state needs only to cease what it is doing for the right to be fulfilled, whereas in the latter case the state has to undertake positive activities to fulfil the right, which can be more problematic in various ways. In particular, the environmental protection programme may require the diversion of resources to its accomplishment". Hayward, 2000, p. 150.

75   El-Hinnawi–a Professor of the National Research Centre in Cairo–was the first who used this term. In a booklet written for the United Nations environmental programme in 1985, El-Hinnawi defined environmental refugees as those people: "who have been forced to leave their traditional habitat, temporarily or permanently, because of a marked environmental disruption (natural and/or triggered by people) that jeopardized their existence and/or seriously affected the quality of their life". He says environmental disruption refers to "any physical, chemical and/or biological changes in the ecosystem (or the resource base) that render it, temporarily or permanently, unsuitable to support human life",El-Hinnawi, 1985, p. 4. In his report, El-Hinnawi described three major types of environmental refugees: 1) those temporarily dislocated due to disasters, whether natural or anthropogenic; 2) those permanently displaced due to drastic environmental changes, such as the construction of dams; and 3) those who migrate based on the gradual deterioration of environmental conditions. As an additional but smaller category, he included those people who were displaced by the destruction of their environment as an act of warfare. Bates, 2002, p. 465.

Since El-Hinnawi's definition of environmental refugees more than 20 years ago, the concept and many features of refugees have changed over time. Nevertheless, this definition remains the most-quoted because of a number of reasons. Firstly, "it clearly confirms that environmental disruption is a reason (besides wars and social conflicts) for the mass movement of humans in contemporary time. Secondly, the definition includes man-made ecological disasters and thus helps to identify those who are responsible for the related environmental changes. Lastly, the definition does not specify that one should leave his or her country in order to be recognized as an environmental refugee" (Boon, & Le Tran, 2007, p. 89).

The literature that developed after this seminal report has made a distinction based on criteria related to the characteristics of the environmental disruption (Tan Yan, Wang Yi Qian, 2004, p. 613). Bates distinguished between the three forms: disasters, expropriations, and deterioration. First, disaster refugees originate in acute events that are not designed to produce migration. These may be divided between those events caused by natural events and those caused by technological accidents". The second category of environmental refugees involves according to Bates: "the permanent displacement of people whose habitat is appropriated for land use incompatible with their continued residence. Such refugees are usually permanently relocated, sometimes with aid from the group expropriating their land". This situation results from an anthropogenic, acute expropriation of an ecosystem that intentionally dislocates a target population. People forced to leave

though international environmental law provides neither protection nor rights to environmental refugees, one might expect this area of law to evolve and take into account their situation, their vulnerability and their need of international protection through a more human rights based approach to environmental law.

This issue is not the subject of this book since it encompasses several complex matters which cannot be explored here. At various times in history individuals and groups have been compelled to abandon their home state, because of the fear of persecution, occasioned by policies based on religion, race, nationality, social, or political programmes. Environmental disasters could now be added to the above list.[76]

In fact, it has been estimated that 25 million environmental refugees are on the move worldwide due to environmental problems, 50 million are left homeless by cyclones, floods and earthquakes, 90 million are displaced by infrastructural projects. These figures are expected to increase sharply in the next few decades due to the impacts of global warming and the consequences of rising sea levels.

Hence, the term "environmental refugees" describes a new kind of mass human casualty caused by negative ecological impacts.**77** Yet, in spite of the increasing number of environmental refugees, the international community has not suggested any durable solutions (Pathak, 1992, p. 235-236; Cancado Trindade, 1992, p. 244). If environmental refugees could invoke a right to a decent, healthy or viable environment, this would create a substantive environmental right which involves the promotion of a certain level of environmental quality to give protection to those people.

1.1.2 Procedural Environmental Rights

It is worth noting that the absence of agreement among states with regard to the pronouncement of a substantive right to environment, has led scholars to consider human rights in a more instrumental approach, and to grant substance to environmental rights by identifying those rights, the enjoyment of which

their residences as land is appropriated for development constitute the first sub-type of expropriation refugees. The third type of environmental refugee according to the author "involves people affected by the gradual deterioration caused by anthropogenic alteration of their environment. Migration that stems from deterioration is not planned, even though the disruption of the environment may be quite deliberate".

76 Intergovernmental Panel on Climate Change (IPCC) 2008a; IPCC. Intergovernmental Panel on Climate Change (IPCC) 2008b; A. Williams, 2008, p. 502.

77 See the literature relating to the Environment Refugees: Bates, 2002, p. 465; Hugo, 1996, p. 105; Myers, 1997, p. 167; Myers, 1993, p. 752; Ramlogan, 1996, p. 81; Suhrke, 1994, p. 473; Westing, 1994, p. 110; Westing, 1992, p. 201.

could be considered a precondition for useful environmental protection.[78] In particular, they focused on the procedural right to environmental information, public participation in decision-making and remedies in the event of environmentaldamage.

The contemporary conception of procedural rights can be traced back to the 1948 Universal Declaration of Human Rights, which stipulated that citizens should be provided with instruments permitting them to voice their opinions in decisions affecting them,[79] to participate in the decision-making process[80] and to have the possibility of redress in cases where decisions have impinged their rights.[81]

These rights are also very important in the field of environmental law and it has been argued that they represent "the pivot in a trilateral relationship of individual/human rights, democracy and environmental protection" (Handl, 1992, p. 139-40). Indeed, they have seen participation "as vital to the flourishing of democracy"[82] and as a requisite to a democratic citizenship.

In fact, the effectiveness of any substantive environmental right presupposes the establishment of a wide range of environmental procedural rights, which would facilitate the practice of environmental citizenship.[83] These rights play "an important role in safeguarding the environment because people contribute to environmental decline, therefore their active participation must be required to prevent it" (Douglas-Scott, 1996, p. 128).

Some scholars have held that procedural rights are more important for the protection of the planet than substantive rights. According to Boyle, Douglas-Schott, Cameron and Mackenzie, "effective environmental rights should be principally procedural in character" (Anderson, 1996, p. 1).

Numerous arguments in favour of this position can be identified. First, the individuals who make the decisions are the same as those who pay for the results of the decisions. Second, as already seen above, it is very complicated to reach a unique accurate formulation of a substantive right to a decent environment

---

78 See e.g. Kiss, 1976, p. 9-15, p. 445.

79 Art. 19.

80 Art. 21.

81 Art. 8.

82 Douglas-Scott, 1996, p. 113. Moreover, "a procedural or participatory approach promises environmental protection essentially by way of democracy and informed debate". Pateman, 1970, p. 43; Birch, 1993. Kelsen's view is: "Political freedom, that is, freedom under social order, is self-determination of the individual by participation in the creation of social order". Kelsen, 1961, p. 285.

83 "In short, a procedural approach may be justified as attempting to provide environmental protection by way of democracy": see Douglas-Scott, 1996, p. 113.

since the quality of the environment is a value judgment, which is too complex to codify in legal words, and which will differ "across cultures and communities" (Anderson, 1996, p. 1).

Then, the emphasis on procedural rights helps not only shape domestic environmental policies, but it might also smoothen the progress of resolving trans-boundary environmental policies, as well as disputes.

Finally, this regime aimed at increasing transparency and accountability in the management of environmental matters by considering that environmental awareness is an important step towards the construction of an "environmental citizenship" as well as an ecological citizenship, which, like the environment, is not attached to national boundaries.

1.1.2.1 Access to Environmental Information

It has been emphasised by Holder: "Access to information is a crucial element of a democratic society, a precondition to basic rights to vote or free speech, and certainly of any form of participation in decision making" (Holder, & Lee, 2007, p. 100). Further, "information generates knowledge and knowledge generates power" (Krämer, 2004, p. 1). Indeed, only a citizen empowered by information, *e.g.* engaged in environmental choices and aware, may make public authorities accountable for their policy choices and, thus, voice her/his concerns to defend the environment.

A right to information can indicate, narrowly, freedom to look for information, or, widely, a right of access to information, or even a right to obtain it. The state has the duty to refrain from interfering with public action to acquire information from the authorities of the state or private bodies. A further duty of the state is to disseminate all relevant information concerning both public and private plans and projects that might have an impact on the environment (Shelton, 2006, p. 26).

Recently, it has been suggested that access to government information should be considered a fundamental human right, for the reason that knowledge of the activities of one's leaders is being seen as "crucial to the maintenance of other human rights" (Cramer, 2009, p. 79; Birkin-Shaw, 2006, p. 177-179).

The underlying idea is that citizens have the right to be acquainted with the methods of how authority power is exercised on their behalf, and therefore access to information can contribute to a better involvement of the citizens in the tasks of their governments. Thus, freedom of information is principally understood as a fundamental right of environmental citizenship and at same time this aspect is also linked to ecological citizenship because access to government information creates a sense of responsibility *vis-à-vis* the environment among citizens to act as "watchdogs over their leaders" when these take a decision concerning the environment (Cramer, 2009, p. 73).

Moreover, access to information also has the opportunity of making the environmental decision-making process a more democratic and efficient process. For this reason, access to information is an essential requirement to exercise participation rights.[84]

1.1.2.2 Participation in Environmental Protection

The most important role played by environmental citizens in environmental protection is participation, which may be political and administrative participation in decision-making. The origin of public participation is "the right of those who may be affected, including foreign citizens and residents, to have a say in the determination of their environmental future" (Shelton, 2006, p. 26).

In fact if a decision is made with the advice and participation of several people, it can reduce the possibility of making a decision which could damage the environment and therefore it could avoid restoration.[85] Public participation processes have been emerging in the policies and environmental regulations of some states since the late 1960s and 1970s. This phenomenon coincided

---

84   It is worth noting that some authors talking about participation include also proceduralisation: "A complementary starting point for proceduralisation is that many vital environmental (and other) decisions in modern states are taken in non-majoritarian institutions, which are not subject to either direct or indirect democratic control, and are not simply applying clear rules that do have democratic approval. There are various reasons for delegating decision-making in this way, the most common justification in the environmental field being the need for detailed technical expertise. There is then a tension between that need for independence and the control of decision-making. Lee, 2003, p. 204.

85   Indeed, *restoration* has been considered negative by different scholars. Elliot argued that ecological restoration, the practice of restoring damaged ecosystems, was akin to art forgery. Just as a copied art work could not reproduce the value of the original, restored nature could not reproduce the value of original nature, conceived as a form of non-anthropocentric and intrinsic, as opposed to merely instrumental, value. For those unfamiliar with the literature in environmental ethics, intrinsic value in this subfield of philosophy is often taken to mean the worth objects have in their own right, independent of their value to any other end and instrumental value is, broadly speaking, the worth objects have in fulfilling other ends. For many environmental ethicists the principle goal of an environmental ethics is to describe the intrinsic (or inherent) value of nature as opposed to its merely instrumental value for human use and consumption. Once an account of the intrinsic value of nature is found then perhaps a range of moral obligations can be derived for things having that value. Many environmental ethicists see the valuing of nature as only instrumental to human ends as part of the cause of human disregard for the environment, and for today's environmental problems. Many environmental philosophers assume that a non-anthropocentric account of natural value is needed to reject instrumental valuing of nature and so any environmental ethic must endorse some kind of intrinsic value account of the value of nature (or at least a non- instrumental account). In turn, it is thought that anthropocentrists can only value nature in instrumental terms. See for example much of the work of Holmes Rolston III, & Baird Callicott on this point. Elliot, 1982, p. 81.

with political disturbances around the world when civil society started to ask for more democratic governance and environmental protection.[86]

From the period of the 1970s to the early 1980s, doctrine and critics have highlighted the importance of citizens to achieve economic development in an environmental manner (Spyke, 1999, p. 263). Consequently, during the 1990s, consultation and participation turned into the buzzwords of environmental decision-making, feeding into broader discourses on "good governance" (Steffek, & Nanz, 2005) "environmental justice" and "environmental citizenship" (Richardson, & Razzaque, 2006, p. 168).

Today the involvement of citizens in environmental decision-making processes has been rationalised from a procedural and a substantive perspective (Richardson, & Razzaque, 2006, p. 170). The latter is based on arguments that the public is required to participate in solutions as well as in decisions (Lee, & Abbot, 2003, p. 83). The former is found in the fact that it provides democratic legitimacy of those decisions.

With regard to both perspectives, several schools of thought on the rationale and role of public participation have arisen. First, there is the *rational elitism* school, which treats environmental policy as complex and technical. It emphasises decision-making by experts, and concedes limited participation to the general public when they hold information that may assist experts (Barton, 2002, p. 84).

Another strand of the *rational elitism* model is known as *corporatism*. Corporatist modes offer "only a functional representation to representatives of large strategic groups such as trade unions, industry and business councils, and sometimes renowned environmental NGOs" (Schmitter, & Lehmbruch, 1979).

A second approach to participation is the liberal democratic one, which emphasises the procedural rights of individuals and NGOs to be consulted and heard in decision-making (Habermas, 1973). This model can be also limited by procedural reforms: citizens may be heard, but their views are taken into account in discretionary decision-making. These potential restrictions have encouraged a third model of participation, called deliberative democracy, which gives concrete decision-making power to citizens and re-orients decision processes to include fundamental ethical and social values (Richardson, & Razzaque, 2006, p. 166).

The above mentioned approaches to participation are not mutually exclusive and elements of each have characterised many multilateral and bilateral

---

86 *E.g.*, in UK, in its planning legislation of the 1960s. the creation of the Royal Commission on environmental Pollution, 1969, and the Department of the Environment, 1970, was the governmental response to these public pressures: McCormick, 1995.

agreements which refer to or guarantee public participation.[87] For example, in the Climate Change Convention, Article 4(1)(i) obliges Parties to promote public awareness and to "encourage the widest participation in this process, including that of nongovernmental organisations" (Shelton, 2006, p. 26).

An important improvement coming from public participation is that more effective environmental protection through participation in decision-making might involve a change in individuals' behaviour. Thus, this character is more linked to ecological citizenship. For instance, concerning the problems raised by waste, solutions have to focus not only on industrial waste production or on waste management, but also on the active involvement of individuals, since they will be asked to reduce waste production, to separate waste for reuse or recycling, to increase composting, and to accept waste management facilities near their homes. Thus, participation does not just have to be linked to the rights approach but also to the duty approach. In fact, participation in the private sphere is fundamental for the effectiveness of environmental policy. In this regard, it has to be remarked that "there is frequently a didactic or at least awareness-raising element to environmental participation democracy" (Lee, 2003, p. 203).

1.1.2.3 Access to Justice in Environmental Matters

Access to justice encompasses three fundamental aspects which will be examined in detail later: first, review procedures with respect to information requests; then, review procedures with respect to specific decisions which are subject to public participation requirements; and finally, challenges to breaches of the right to environment and also environmental law in general.

Opening up access to justice in the environmental field to members of the public is a democratic necessity given that the implementation and enforcement

---

[87] Protocol to the 1979 Convention on Long-Range transboundary Air Pollution Concerning the Control of Emissions of Volatile Organic Compounds or Their transboundary Fluxes, Article 2(3)(a)(4), Nov. 18, 1991, 31 I.L.M 568; Convention on the Protection and Use of transboundary Watercourses and International Lakes, Article 16, Mar. 17, 1992, 31 I.L.M. 1312; Convention on the transboundary Effects of Industrial Accidents, Article 9, Mar. 17, 1982, 2105 U.N.T.S. 460; Convention for the Protection of the Marine Environment of the North-East Atlantic, Article 9, Sept. 22, 1992, 32 I.L.M. 1072; Convention on Civil Liability for Damage Resulting from Activities Dangerous to the Environment, arts. 13-16, June 21, 1993, 32 I.L.M. 1228; North American Agreement on environmental Cooperation, arts. 2(1)(a), 14, Sept. 14, 1993, 32 I.L.M. 1480; Danube Convention, at Article 14; Protocol Concerning Specially Protected Areas and Biological Diversity in the Mediterranean, Article 19, June 10, 1995, 1999 O.J. (L 322) 3; Joint Communique and Declaration on the Establishment of the Arctic Council, pmbl, arts. 1(a), 2, 3(c); Sept. 19, 1996, 35 I.L.M. 1382; Kyoto Protocol to the United Nations Framework Convention on Climate Change, Article 6(3), Dec. 11, 1997, 37 I.L.M. 22; Stockholm Convention on Persistent Organic Pollutants, Article 10(1)(d), Sept. 22, 2001, 40, I.L.M. 532.

of environmental protection laws is a task which governments alone cannot fully accomplish: in a democratic society based on the rule of law, Hayward has argued that "individual citizens and their various associations have a role to play in this field too" (Hayward, 2000, p. 143).

Yet, in the book "Defending the Environment, A Handbook for Citizen Action" Sax suggests that citizens' initiatives in court in the environmental field have presented also other advantages. First, the court is attractive because "free of the constraints which familiarity and close dealing tend to breed, it can bring fresh insights to problems of environmental management".

Second, the courts are "not to be used as substitutes for the legislative process [...] but as a means of providing access to legislatures so that the theoretical processes of democracy can be made to work more effectively in practice". Furthermore, these means can be "used to bring important matters to legislative attention, to force them upon the agendas of reluctant and busy representatives".

The author concludes "if we are to save the environment, rather than merely revere it the citizen can no longer be put off with the easy advice to go get a statute enacted or "wait until election day, while the bulldozer of chain saw stands ready to move" (Sax, 1972, p. XVIII-XIX).

1.1.3 Criticism of Environmental Rights

1.1.3.1 Anthropocentric Approach to Environmental Rights

The anthropocentric approaches to environmental protection are seen as perpetuating the values and attitudes that are at the root of environmental degradation. There is real concern among many commentators over the inherent anthropocentricity of environmental human rights. This one-sided approach to environmental rights reinforces the idea that the environment exists only for human benefit and has no intrinsic worth.[88]

This approach deprives the environment of direct and comprehensive protection, as human life, health and standards of living are likely to be the aims of environmental protection. Indeed, according to Taylor, the environment is only "protected as a consequence of, and to the extent necessary to meet, the need to protect human wellbeing. An environmental right thus subjugates all

---

88 Birnie and Boyle point out that "by looking at the problem (of anthropocentric human right) in moral isolation from other species, such a right may reinforce the assumption that the environment and its natural resources exist only for human benefit, and have no intrinsic worth in themselves".

other needs, interests and values of nature to those of humanity. Environmental degradation or loss of ecological integrity as such is not sufficient cause for complaint; it must be linked to human wellbeing". And consequently, the individual has the right to initiate legal action and there is "no guarantee of its utilisation for the benefit of the environment, nor is there any recognition of nature as the victim of degradation" (P. Taylor, 2009, p. 99).

Hence, environmental rights result in "creating a hierarchy" where humanity has a superior position, separate from other species or of the planet. Protection stems from human-centred environmental rights and so the actual state of the environment is determined by the needs of humanity, not the needs of other members of the natural community (Bosselmann, 2008, p. 127).

Despite the criticism of the anthropocentric approach, a number of arguments mitigate these concerns. According to Bosselmann, "First, it is suggested that a degree of anthropocentrism is a necessary part of environmental protection. Not in the sense of humanity as the centre of the biosphere, but because humanity is the only species, that we know of, which has the consciousness to recognise and respect the morality of rights and because human beings are themselves an integral part of nature" (Bosselmann, 2008, p. 128).

Briefly, the interests and duties of humanity are inseparable from environmental protection. Thus far, Shelton agrees (Shelton, & Memon 2002, p. 8-9), but goes on to argue that an environmental human right could be complementary to wider protection of the biosphere which recognises the intrinsic values of nature, independent of human needs.

Birnie and Boyle point out that the anthropocentric approach may reinforce the assumption that the environment exists only for human benefit; nevertheless, participatory environmental rights may integrate the human rights claims within a broader decision-making framework capable of taking into account, amongst other factors, intrinsic values, the needs of the environment and the needs of present and future generations.

Rolston (Rolston, 1993, p. 251) also advocates a compromise position. He accepts the paradigm of human rights for the protection of the human need for environmental integrity, but in addition suggests the elaboration of human responsibilities for nature (Bosselmann, 2008, p. 128). Consequently, a need to integrate a new ecocentric approach is growing: the interpretation of environmental rights should change and be tempered by ecological responsibilities (Bosselmann, 2008, p. 319).

1.1.3.2 Introduction of a New Ecocentric Approach to Environmental Rights

According to the ecological approach, when formulating an environmental human right humans should be viewed as a unit in the ecological system and one "should proceed on the basis that his environmental rights are qualified by the rights and interests of other affected sectors of the ecology" (Pathak, 1992, p. 223).

In fact, man does not enjoy a higher position on the tree of evolution over the rest of nature but "he is, indeed, merely a component equal with the other components of the ecological bio-system" and he does not hold a superior status but just a different kind of status which grants him responsibilities towards other species and the planet. Consequently, Man has the duty to articulate and defend the rights of other occupants of the planet. In adopting this approach, one does not seek to view the environment in a "homocentric dimension" (Pathak, 1992, p. 205-206), but the perspective is overturned and humanity is an integral part of the biosphere, nature has an intrinsic value and humanity has obligations towards nature. In order to move towards this new approach, Bosselmann has suggested that the introduction of "ecological limitations, together with corollary obligations, should be part integrate to the environmental rights discourse" (Bosselmann, 2008, p. 130).

Of course in this perspective, it can be argued that if environmental rights were to have the capacity to "trump" (Anderson, 1996, p. 21) other interests, they could compromise a 'right to development' for example, or just simply a developing nation's efforts to satisfy the economic, social and cultural rights of its peoples. For this reason the decisions relating to the environment often require a "balancing of interests" (Turner, 2009, p. 52).

Following this last point of view, Waite has suggested focusing on a "polycentric" approach (Waite, 2007, p. 395), taking into account the interests of human beings, individual non-humans and the environment as a whole. A polycentric approach to environmental protection "not only emphasises the convergence of interest between humankind and the environment; it also ensures a balancing of those interests where they conflict" (Waite, 2007, p. 395).

In conclusion, it may be said that the ecological approach to human rights has to acknowledge the interdependence of rights and duties. Cullet, discussing the implementation of environmental rights, states: 'the only way to achieve an effective implementation of the rights is to lay a duty on the holders of the rights, to participate in the enhancement of the environment" (Cullet, 1995, p.25).

Human beings need to use natural resources, but they also completely depend on the natural environment. This makes self-restrictions and recognition of the duty approach essential, not only in practical terms, but also in normative terms. Entitlements to natural resources and a healthy environment, usefully expressed as rights, can be integrated by duties which respect and guarantee ecological boundaries. These duties can be expressed in ethical and legal terms as they define content and limitations of human rights (Bosselmann, 2008, p.146).

To conclude, it is worth noting that environmental rights are not a magic formula that will cure the planet of its environmental sickness. There will need to be awareness in society and effective education about the causes of environmental destruction together with a social conscience that action on the part of all groups and individuals is required and most important, that environmental rights must be completed by ecological duties (Horn, 2004, p.268).

2 Ecological Citizenship

The original meaning of the term "ecological citizen"[89] is "citizen of the world", rather than with regard to a particular *polis*, nation, or bioregion.[90] Several theorists have looked at the role of obligations in citizenship in an attempt to identify agents for the transformation of existing socio-ecological orders (J. Barry, 2002, p. 133).

Saiz asserts that 'ecological citizenship is still "under construction", but it can already be seen that this has its own architectural inflections that break with traditional notions of citizenship'. As such, the ecological citizen must be constituted in a new political space that overflows the boundaries of discrete nation states (Latta, 2007, p. 381). Dobson gives ecological citizenship a totally new context as a type of post-cosmopolitan citizenship (Caney, & Simons, 2005, p. 747).

As seen above, this citizenship differs from the republican citizenship because it is a non-territorial form of citizenship, due to the fact that it extends beyond territorial boundaries, and second, because it embraces both the private and public sphere (Dobson, 2003, p. 82).[91] Concerning the first characteristic, it is worth noting that the dimension inside which citizens operate is the planet as a whole. This is especially due to the circumstance that numerous environmental problems are trans- or international in scale (Dobson, & Bell, 2006, p. 5-6).

---

89 Bell, 2005, p. 179; Christoff, 1996; Clarke, 1999; Dean, 2001, p. 490; Drevensek, 2005, p. 226; Luque, 2005, p. 211-225; Sáiz, 2005, p. 163; Seyfang, 2005, p. 290; G. Smith, 2005, p. 273; M.J. Smith, 1998; Stephenson, 1978, p. 21; Thomas, & Twyman, 2005, p. 115.

90 The first conceptualisation of this citizenship is from a article in Dobson, 2000; Dobson, 2003, p. 67; J. Barry, 2006, p. 21.

91 Dobson: Thus the typical characteristics of the ecological, also post-cosmopolitan, citizenship are the "non-reciprocal nature of the obligations associated with it, the non-territorial yet material nature of its sense of political space, its recognition that this political space should include the private as well as the public realm, its focus on virtue and its determination to countenance the possibility of private virtues being virtues of citizenship"; Melo-Escrihuela, 2008, p. 113.

The second characteristic of republican citizenship is the emphasis on obligations and responsibilities rather than on rights[92] in the private and public sphere. The idea is also that on the one hand, in the public sphere ecological problems do not get solved without participation and without "virtuous citizens checking their government, stimulating it", and on the other hand, in the private sphere these problems do not get solved "without popular support" (Wissenburg, 2004, p. 73).

Indeed if flights are cheap, people will fly; if gas is cheap, people will drive. Most citizens, most of the time, will act only in response to external motivations of price, punishment, or prohibition. For this reason the use of legal or economic instruments is therefore a necessary part of the environmentally sustainable whole.[93]

Nevertheless the ecological individual should also act in a way that is not always in their best interest independently from the legal rules (Dobson, 2009, p. 18).

In contrast to environmental citizenship, which focuses solely on the environment, ecological citizenship aspires to the promotion of global and environmental justice.[94] In fact, ecological citizenship diverges from

---

92   It must be noted that the individuals have those rights and responsibilities ,as residents of planet Earth' vis-à-vis the future generations, as we will see in detail later, and Nature. Draft declarations of human responsibilities such as the Earth Charter focus on duties toward the environment. See The Earth Charter, princs. pp. 4-5, Mar. 2000, available at www.earthcharter.org/files/charter/charter. pdf (encouraging the protection and restoration of ecological systems and taking action to prevent future environmental harm). Many proponents of this approach posit ecological rights or rights of nature as a construct to balance human rights, attempting to introduce ecological limitations on human rights. "The objective of these limitations is to implement an eco-centric ethic in a manner which imposes responsibilities and duties upon humankind to take intrinsic values and the interests of the natural community into account when exercising its human rights. P. Taylor, 1998, p. 309-310; Shelton, 2006, p. 26; Mank, 1996, p. 445.

93   Flynn, Bellaby, & Riccim 2008, at p. 771. They underline "The problems of breaking out of an economy totally dominated by fossil fuels, and difficulties in bringing about major changes in people's lifestyles towards sustainability, have been noted by many other commentators (Murphy, & Cohen 2001, p. 225). One important recent report on moves towards sustainable consumption and social justice described the current inertia, expressed in a prevailing public attitude of "I will if you will" (Sustainable Development Commission 2006). However, at the 'micro' level, the apparent gap between attitude and action cannot be explained solely in terms of lack of information. Hobson showed in a detailed qualitative study of household consumption and lifestyles, that the limits on people's willingness to change is partly linked with specific discourses or rhetoric about consumption and the environment, and partly to do with the deeply-embedded nature of everyday practices".(Hobson, 2001, p. 19; Hobson, 2003, p. 95)

94   Environmental justice is defined as the "fair treatment and meaningful involvement of all people regardless of race, color, national origin, or income with respect to the development, implementation and enforcement of environmental laws, regulations and policies. The environment justice framework rests on developing tools and strategies to

environmental citizenship in that the former foresees a different society that is not only sustainable but also just where the fulfilment of duties is a way of assuring justice (Melo-Escrihuela, 2008, p.113).

According to Dobson the obligation to seek for more just arrangements falls upon those who currently enjoy more than their share of the world's resources, and it is an obligation that arises from the simple fact of such unjust distributions (Latta, 2007, p. 377). Dobson introduces the notion of footprints: the ecological citizen wants to make certain that her/his "ecological footprint"[95] does not harm the capacity of present and future generations to "pursue activities important for their well-being" (Dobson, 1998, p. 119). Consequently, global injustice in the shape of disparate footprints and unbalanced power relations represents the historical situation on which ecological citizenship is built. It generates obligations for citizens to act so as to remedy these injustices.

It has also been underlined by Christoff that the role of the ecological citizen, defined as "*homo ecologicus*", is "to defend the rights of future generations and other species just as we are morally obliged" (Christoff, 1996, p. 159). This means that humans "must assume responsibility for the future humans and other species and "represent" their interest and potential choices according to the duties of environmental stewardship" (Hay, 2002). Thus, there are two fundamental obligations, one to present and future generations, and another to nature.

Although ecological citizenship has a cosmopolitan and non-territorial character that does not mean that the role of the state has been eliminated. Its role remains an exceptionally important focus on ecological citizenship; it can provide the legal and material support for further ecological democratisation (Christoff, 1996, p. 151). Indeed, the dominant position is that the transition toward ecological citizenship requires governmental policies to create the conditions and spaces for its exercise.[96] It is worthy noting that, from a legal –

---

eliminate unfair, unjust, and inequitable conditions and decision", Bullard, 1996. See also: G. S. Johnson, 2009, p. 17.

95 The term "ecological footprint" comes from Wackernagel, & Rees, 1996. The footprint size "is arrived at by dividing the total land available, and its productive capacity, by the number of people on the planet, and the figure usually arrived at is somewhere between 1.5 and 1.7 hectares. Inevitably, some people have a bigger impact – a bigger footprint – than others". Dobson, 2004, p. 122.

96 Wolf remarks some critiques to this theory "In critiques of Dobson's work, two crucial weaknesses of ecological citizenship theory, and its potential application in practice, have been identified. First, as Saíz (Saíz, 2005, p. 163) puts it, "Dobson's insistence on the efficacy of individual political agency" (p. 176) is a critical point of weakness because it implies that individuals can be relied upon to strive to be better citizens. This ignores that individuals act within a social, economic, cultural and institutional context that shapes and constrains citizens' ability to act in particular ways. A related point is made also by Luque

and not just philosophical–point of view, the Aarhus Convention is an example of how ecological duty can be incorporated and how it can be become also a legal obligation.[97]

2.1 Ecological Duties

As a counterbalance to the rights-based approach which offers only indirect and limited ecological protection and reinforces the anthropocentric value system that is at the root of ecological degradation, there is an additional view. What is necessary is more emphasis upon the adoption and exercise of responsibilities towards all life, including non-human life (P. Taylor, 2009, p. 89), and a special responsibility to "care for the planet" (Weiss, 1990, p. 199).[98]

Increasingly, it is being pointed out that in many cultures individuals have duties and responsibilities towards others and the wider community. Traditionally,

---

(Luque, 2005, p. 211) who points out that the ecological footprint metaphor used by Dobson implies that individuals who recognise their footprints to be too large can satisfy their responsibility to those impacted by simply reducing the size of their footprints. But "unless 'doing one's share' focuses most of all on bringing about structural change, the deactivation potential of the ecological footprint metaphor would be of concern". In addition, it should be added that reducing an individual's footprint in a developed country does not necessarily enable access of an individual in a developing country to any resources. In this case, even when focusing on bringing about structural change, it would be extremely difficult to effect changes at the global and national economic and institutional scales which could allow such access by those currently disadvantaged". Second, changes in individual's impacts may not be sufficiently large. This point is recognised by Valdivielso (Valdivielso, 2005, p. 239) who suggests that many motivated ecological activists do not have the opportunity to maintain sustainable consumption. Living in the developed world often means adhering to a minimum living standard that embodies a lifestyle intricately intertwined with patterns of consumption, which in turn are culturally and socially embedded needs of mobility, food, work, housing, training or leisure. As a result, in these cases "the least possible impact is often still much higher than the desired impact". The above discussion shows that ecological citizenship is a young and largely theoretical concept. Despite the criticisms outlined, the concept may have explanatory power in areas of human motivation and behaviour that extends beyond that of other approaches". Wolf, 2007, Helsinki.

97 The Preamble of the Aarhus Convention recognises that "every person has the right to live in an environment adequate to his or her health and well-being, and the duty, both individually and in association with others, to protect and *improve* the environment for the benefit of present and future generations".

98 See also Weiss, 1989; 1992a, p. 385. Moreover the fulfilment of intergenerational obligations requires attention to certain aspects of intergenerational equity "When future generations become living generations, they have certain rights and obligations to use and care for the planet that they can enforce against one another. Were it otherwise, members of one generation could allocate the benefits of the world's resources to some communities and the burdens of caring for it to others and still potentially claim on balance to have satisfied principles of equity among generations".

the duty-approach offers a subordinated prospective. Nevertheless, during the French Revolution the idea that citizenship is more about duties towards the Republic than rights was dominant.

The slogan "no rights without responsibilities" is starting to take a new position in modern green political thought. Indeed, the other face of environmental rights presumes an active attitude on behalf of citizens, and even more, a citizens' duty to protect the environment. Each person has the right to have his or her environment protected, but is also obliged to contribute to the common effort. Citizens are not passive beneficiaries, but share responsibilities on the formation of all community interests (Kiss, 1992, p. 201).

For this reason, some scholars have recognised that positive ecological duties often "flow from rights".[99] Habermas has suggested in another context to take the next step and establish a legal duty to make active use of democratic rights (Habermas, 1991, p. 1). A rights-based approach could be used to specifically create legal duties for all decision-makers in relation to protection of the environment.[100] Exponents of this viewpoint include Nanda and Pring who state that the right to environment generates duties beyond those required of the government.

The right would also entrain the imposition on individuals, organisations, and corporations of a duty to refrain from activities that harm the environment (Nanda, & Pring, 2003, p. 475). A duty has first to be laid upon all individuals as their combined actions can have a significant impact.

Moreover, it is worth noting that ecological duty has its background in the principle of ecological responsibility. Indeed, ecological responsibility is not a new topic: Jonas was one of the first to propose this principle as a way to cope with the ecological problems generated by technological society.

In *The Imperative of Responsibility* he revives the earlier ethics of virtue from ancient Greek philosophy, criticises human interactions with nature for being based solely on *techné*, observes that ethical principles have not kept up with technological changes, and proposes a new imperative: "Act in such a way that the consequences of your action are compatible with the permanence of genuine human life on Earth" (Jonas, 1979, p. 36).

This principle is necessary because technology and progress have destroyed nature. Jonas argues that this responsibility is motivated by the drive for survival,

---

99   Desgagne, 1995, p. 263. Weiss affirms "Planetary rights and obligations are integrally linked and are in the first order collective obligations and collective rights. The rights are always associated with the obligations": Weiss, 1989, p. 45.

100   Gormley, 1990, p. 85; Nickel, 1993, p. 281. Nickel says that "persons, organisations, and corporations have a duty to refrain from activities that generate unacceptable levels of environmental risk".

insofar as "the birth of each new child gives humanity the perspective to begin anew in face of mortality" (Jonas, 1979, p. 241). Because the survival of the human species is impeded by egoism, destruction of nature, and catastrophes that lead to global crises, Jonas concludes that the principle of responsibility must be introduced (Jonas, 1979, p. 390; Melle, 1998, p. 329).

This conclusion is justified by Jonas: "Responsibility requires not only a power or capability to guide one's own actions but also the recognition of an obligation–which is best seen in the paradigmatic examples of those parents who take care of their children and the politicians that assume responsibility for their citizens. Thus, the care for the "life of others" is the ethical basis upon which he places responsibility" (Nascimento, 2009).

2.2 Two Fundamental Ecological Duties

2.2.1 Duty to Protect the Environment for Present and Future Generations

All decisions taken today will affect the quality of life for generations to come. Indeed, future peoples will suffer from the ways in which the environment is degraded and the extent to which the earth's resources are wasted (Beckman,2007).

Philosophy, religion, green political thought and some legal traditions from diverse cultural traditions have already recognised that man is trustee or steward of the natural environment and from this arises man's duty to conserve the planet for present and future generations.[101] Nevertheless this recognition

---

101 "There are roots in the common and the civil law traditions, in Islamic law, in African customary law, and in Asian non-theistic traditions. The proposed theory of intergenerational equity finds deed roots in the Islamic attitude toward the relation between man and nature". (Islamic Principles for the Conservation of the Natural Environment, 13-14 (IUCN and Saudi Arabia 1983). Islamic law gives to man all the resources of life and nature; each generation is entitled to use the resources but must care for them and pass them to future generations. "In the Judeo-Christian tradition, God gave the earth to his people and their offspring as an everlasting possession, to be cared for and passed on to each generation (Genesis 1:1-31, 17: 7-8 "I will maintain my Covenant between Me and you, and your offspring to come, as an everlasting covenant throughout the ages, to be God to you and to your offspring to come. I give the land you sojourn in to you and to your offspring to come all the land of Caan, as an everlasting possession. I will be their God".) This has been carried forward in both the common law and the civil law tradition. The English philosopher John Locke, for example, asserts that, whether by the dictates of natural reason or by God's gift "to Adam and his posterity", mankind holds the world in common. Man may only appropriate as much as leaves enough, and as good

is not universal and almost all environmental theories note there is a huge lack of inter-generational and intra-generational equity. First, political leaders fail to adequately consider future interests in evaluating policy options. But "this myopia" is not the outcome of a lack of a concern for children or future inhabitants of Earth; instead, it is "the result of institutional constraints that encourage political leaders to prioritise the short-term needs of voters" (Wolfe, 2008, p. 1897).

2.2.1.1. Intergenerational Equity: Duty Vis-à-vis Present Generation

Intergenerational equity concerns the lack of adequate consideration by political leaders in the developed countries vis-à-vis present generations living in developing countries. Intergenerational equity is narrowly linked to the footprint discourse, e.g. the inequity of the share of the planet's natural resources among members of the present generation.

---

for others. He has an obligation not to take more fruits of nature than he can use, so that they do not spoil and become unavailable to someone else – e.g., an obligation not to waste the fruit of nature (Locke, 1968). To be sure, there are many instances where law has been used to authorise the destruction of our environment, but the basic thesis that we are trustees or stewards of our planet is deeply imbedded. In the civil law tradition, this recognition of the community interest in natural property appears in Germany in the form of social obligations that are inherent in the ownership of private property (Dozer, 1976). Rights of ownership can be limited for the public good, without the necessity to provide compensation to the owners. Thus legislatures can ban the disposal of toxic wastes in ecologically sensitive areas and invoke the social obligation inherent in property to avoid monetary compensation to the owner of the land. In common law countries such as the United States, local governments can do this through the exercise of the police power- the power to protect the health and welfare of its citizens – or the public trust doctrine. The social legal tradition also has tools which recognise that we are only stewards of the earth. Karl Marx states that all communities [...] are only possessors or users of the earth, not owners, with an obligation to protect the earth for future generations. According to African customary law we are only tenants on Earth with obligations to past and future generations. Under the principles of customary land law Ghana, land is owned by a community that goes on from one generation to the next. A distinguished Ghanaian chief said " I conceive that land belongs to a vast family of whom many are dead, a few are living, and a countless host are still unborn" (Ollennu,1962). Land thus belongs to the community, not to the individual. The Chief of the community or head of the family is like a trustee who holds it for the use of the community. Members of the community can use the property, but cannot alienate it. Customary laws and practices of other African communities, and indeed of peoples in other areas of the world, also view natural resources as held in common with the community promoting responsible stewardship and imposing restrictions on rights of use (Blanc-Jouvan , 1971). The non theistic traditions of Asia and South Asia, such as Shintoism, also stress a respect for nature and our responsibilities to future generations as stewards of this planet, in most instances they call for living in harmony with nature. Moreover, Hinduism, Buddhism and Jainism indirectly support the conservation of our diverse cultural resource in their acceptance of the legitimacy of other religious groups. Weiss, 1989.

As said above, Dobson's ecological citizenship emerges as an obligation to correct the injustices inherent in the material relationships encompassed by the notion of an ecological footprint. Chambers point out that "every organism, be it a bacterium, whale or person, has an impact on the earth, we all rely upon the products and services of nature, both to supply us with raw materials and to assimilate our wastes, the impact we have on our environment is related to the quantity of nature that we use to sustain our consumption patterns (Chambers, Simons, & Wackeragel, 2000, p. XIII).

Individuals who currently leave inordinately large ecological footprints are obliged to act by decreasing their consumption of earth's resources (Latta, 2007,p. 377).

Indeed, in Dobson' words "the nature of the obligation is to reduce the occupation of ecological space, where appropriate, and the source of this obligation lies in remedying the potential and actual injustice of appropriating an unjust share of ecological space".

Ecological obligations are "the corollary of a putative environmental right to an equal share of ecological space for everyone [...] those who put down large footprints leave less ecological space for others to inhabit, thereby excluding them from their rightful share of the basic ecological necessities that make a dignified life possible to live" (Dobson, 2004, p. 123).

Consequently, a characteristic of ecological obligations is that they are owed asymmetrically because this duty is borne only by those who occupy ecological space in an unsustainable way so as to compromise the ability of others in present and future generations (Dobson, 2003, p. 82). For instance, global environmental problems, *e.g.* climate change, underline how the impacts of environmental problems in relation to the global level are asymmetric if one puts the attention on the difference between developed and developing countries. Thus, the obligations of ecological citizenship arise from the asymmetric distribution of power and effect between (and among) citizens of developed and developing countries (Wolf, 2007).

2.2.1.2 Intergenerational Equity: Duty Vis-à-vis Future Generation

The intergenerational issue underlines that our responsibilities to future generations demand that we take a long-term perspective (Weiss, 1984, p. 119).

Responsibilities *vis-à-vis* all members of our species exist, as has been well theorised by Weiss in his essay titled *Fairness to Future Generations*. The human holds Earth in trust for future generations. The principle of intergenerational equity forms the basis of a set of intergenerational obligations and rights, or planetary rights and obligations that are held by each generation. According to the author, "when we are born, we inherit a legacy from past generations to enjoy on the condition that we pass it on to future generations to enjoy.

This imposes a set of planetary obligations upon members of each generation and gives them certain planetary rights".[102] Planetary rights and obligations are correlated. They are rights "of each generation to receive the planet in no worse condition that did the previous generation, to inherit comparable diversity in the natural and cultural resources bases, and to have equitable access to the use and benefits of the legacy. This obligation represents in the first instance a moral protection of interests, which must be transformed into legal rights and duties" (Weiss, 1990, p. 203). Consequently, if one generation fails to conserve the planet, at the level of quality received, the "succeeding generation has an obligation to repair this damage, even if it is costly to do so" (Weiss, 1989, p. 21).

It has been underlined by D'Amato that "future generations cannot have rights, because rights exist only when there are identifiable interests, which can only happen if we can identify the individuals who have interests to protect. Since we cannot know who the individuals in the future will be, it is not possible for future generations to have rights" (D'Amato, 1990, p. 190).

This approach presupposes the classic theoretical structure of rights as rights of "identifiable individuals". Nevertheless, planetary, or intergenerational, rights are not rights possessed by individuals. They are, instead, generational rights,

---

102 There are for the authors three categories of Planetary Obligations "First, each generation should be required to conserve the diversity of the natural and cultural resource base, so that it does not unduly restrict the options available to future generations in solving their problems and satisfying their own values, and should also be entitled to diversity comparable to that enjoyed by previous generations. This principle is called 'conservation of options'. Second, each generation *should* be required to maintain the quality of the planet so that it is passed on in no worse condition than that in which it was received, and should also be entitled to planetary quality comparable to that enjoyed by previous generations. This is the principle of 'conservation of quality'. Third, each generation should provide its members with equitable rights of access to the legacy of past generations and should conserve this access for future generations. This is the principle of 'conservation of access'. These three categories of Planetary Obligations are further disarticulated into five duties of use: (I) the duty to conserve resources; (II) the duty to ensure equitable use; (III) the duty to avoid adverse impacts; (IV) the duty to prevent disasters, minimise damage, and provide emergency assistance; and (V) the duty to compensate for environmental harm. The duty to conserve resources requires present generations to conserve both renewable and non-renewable natural resources. Although endangered species and unique natural resources may require strict preservation, this planetary duty generally allows for the sustainable development of resources. The duty to ensure equitable use, defined as 'reasonable, non-discriminatory access to the [planetary] legacy' includes both the duty to refrain from infringing on the access rights of other beneficiaries and the positive obligation to 'assist those who would otherwise be too poor to have reasonable access and use'. The duty to avoid adverse impacts on the environment 'emphasises prevention and mitigation of damage' and includes procedural environmental duties including environ- mental assessment, notice, information and consultation": see Collins, 2007b, p. 321.

which must be conceived of in the temporal context of generations" (Weiss, 1990, p. 198).

Intergenerational planetary rights and duties may be considered as group rights, separate from individual rights; in other words, generations hold their rights and duties as groups in relation to other generations, past, present and future. They exist in spite of the identity of individuals making up each generation. When "held by members of the present generation, they acquire attributes of individual rights, as procedural rights above examined, in the sense that there are identifiable interests of individuals which are protected by the rights" (Weiss, 1990, p. 198).

To sum up, there is an obligation borne by the present generation which involves the protection of the environment for future generations. Thus, each generation is both a custodian and a user of our common natural surroundings. As custodians of this planet, we have certain moral obligations to future generations which we can transform into legally enforceable norms.

2.2.2 Duty to Protect the Environment for the Environment

The second obligation is the duty to protect the environment, *e.g.* the living and non-living creatures. This duty is reflected in the principle of sustainability and cannot be confused with shallow versions of sustainable development. The indispensable element of the new categorical imperative is responsibility for the community of life (Bosselmann, 2008). If translated to political theory, responsibility for all life requires a total rethinking of law and governance.

The key definition of sustainable development, registered in the Brundtland Report to the United Nations in 1987, argued that development and growth were compatible with ecological demands, provided that such development is "sustainable". Moreover, the document states that "sustainable development meets the needs of the present without compromising the ability of future generations to meet their own needs".[103]

This document adds a fictional argument about future situations and generations; however it neglects other environmental aspects. The UNEP report, *Caring for the Earth*, adds that sustainable development aims at "improving the quality of life while living within the carrying capacity of supporting ecosystems".[104] Nevertheless, "sustainable development" still seems to be a contradiction in terms (Attfield, 2003, p. 181), and "the above mentioned problems lead to an emphasis on "sustainability". The strong principle behind

---

103   World Commission On Environment and Development, 1987.
104   UNEP, 1991.

sustainability is the idea of human survival and maintenance of current conditions. It does not necessarily involve, for instance, restoration, revision, or reparation. It means that humans are entitled, for example, to kill other species to provide for food or even to generate riches that aim at maintaining or warranting the survival of future generations. This counterfactual argument opens the door to other claims for justice and solidarity for those who cannot speak for themselves, so that the possibility of arguing for poverty alleviation, animal rights, ecological systems, biotic communities, and natural entities" (Nascimento, 2009).

Consequently, there is a necessity to recognise an obligation of man towards all non-human elements of the planet. The general principle which provides that an obligation arises only upon a correlative right cannot serve here inasmuch as non-human elements cannot be regarded as right-bearing.

Stone (Stone, 1988, p. 56) has suggested that even if non-human, whether animate or inanimate, objects cannot be regarded as possessors of rights, they shall be treated as morally considerable. Moral consideration, he says, creates duties of man towards non-human animate and inanimate objects. The mere fact that "non-human things possess an intrinsic goodness, that is, goods in and of themselves, should be sufficient to attract duties" (Pathak, 1992, p. 225).

2.3 Implementation of Ecological Duties

Given the analysed duties, it is imperative now to develop and implement a strategy for fulfilling ecological responsibilities. Ecological obligations are even more difficult to implement than environmental human rights. The main reason is that they are still at the level of moral obligations despite the fact that they have progressed a few steps towards a transformation into legal duties.

The strategy could encompass the following components: first, codification of obligations and drafting of rules to sanction the violations; then, representation of future generations in decision-making processes; finally, giving a voice to nature, in other words, giving to nature the right to representation (Weiss, 1984, p. 119).

2.3.1 Implementation through Codification

Moral responsibility *vis-à-vis* present and future generations and nature may be differently implemented in law, for instance through a constitutional provision.[105]

---

105 For instance Westra suggests that "eco-crimes represent gross breaches of human

There are a number of ways of achieving this legal implementation. It has been suggested to use international agreements or regional legislations or constitutions, containing provisions for the protection of environmental rights. This could include solemn provisions creating collective and individual responsibilities for the protection and restoration of the ecological basis of all life (Barresi, 1997, p. 3).

The suggestion is not just the codification of ecological duties but also the development of particular regulations that may have the effects of influencing people to change their beliefs and, in turn, to act more sustainably (Davis, 2007; Geisinger, 2002, p. 35; 2009).

Indeed, law can "teach" individuals how to change their behaviour in order to act more sustainably. Traditionally, law does not teach. In fact, traditional views consider law to be an "exogenous force" (MacGregor, 2004, p. 85) which influences an individual by making a desired behaviour either less or more costly to undertake. More recently, however, scholars have begun to consider ways in which law may actually have "endogenous" (MacGregor, 2004, p. 85) force on individuals. In such cases, law affects individuals' beliefs in a way that, even if the legal restraints were removed, the individual would continue to act in accordance with the prior legal command.[106]

---

rights and should be judged accordingly, and no less seriously than (a) attacks against the human person; b) genocide; c) breaches of global justice; and d) crimes against humanity in general": L. Westra, 2004; L. Westra, 2006: "There are proposals to treat major violations of obligations for the conservation of the human environment as international crimes, by labelling these actions as crimes, it can be argued, we would emphasise the common interest of all states and indeed all people in certain planetary values for protecting the human environment for both present and future generations. Any state would then be able to raise claims to enforce the obligations and would have to desist form assisting those who commit the crime. We would be sending an important signal to all members of the international community that certain actions cannot be tolerated". "An alternative approach is to treat actions causing severe and large scale degradations of the human environment as constituting crimes against humanity [...]. There are already several international agreements which arguably make actions causing severe, widespread, and long-term environmental damage war crimes if committed during armed conflicts. Protocol I, adopted in 1977, to the Geneva Conventions of 1949 on armed conflict prohibits "employing methods or means of warfare which are intended or may be expected, to cause widespread, long-term and severe damage to the natural environment. If the Code were to include environmental crimes, individuals could be held responsible, which contrasts with the principles of state responsibility, which only apply to behaviour of states. Indeed the crimes against humanity could be defined to apply only to individuals. This would be an explicit recognition that individuals, whether private or corporate, have an international obligation to humanity as a whole, to present and future generations". Weiss, 1984, p. 89; Weiss, 1989.

106 In this field see, for example: Ellickson, 1991; McAdams, 1997, p. 338; Lessig, 1995, p. 943. See e.g. Posner, 2000, p. 1781; Pildes, & Niemi, 1993, p. 438; Geisinger, 2009.

This last way should surpass the so called "command and control regulation" which consists of a mechanism that includes a wide range of regulatory techniques sharing the basic characteristic that central government regulation dictates a particular end and requires individuals or industry to meet it (Lee, 2002a, p. 114). In fact, a state telling industry and individuals what to do is simply not the most efficient way of achieving social objectives. Thus there are alternative mechanisms whereby financial incentives are used to encourage the desired behaviour.[107] Finally, another way could be to set up an environmental liability regime which could implement the duties and at the same time modify the behaviours.

2.3.2 Implementation through Representation of Future Generations

A measure of how to implement the duty *vis-à-vis* future generations is found in the possibility of *representation* of future generations. As already remarked before, democratic governments have been under extensive criticism for not adequately taking the interest of the unborn into account. In fact, political participation in democracies includes only living people, leaving the "unborn without a voice".[108] Also, the World Commission on Environment and Development reported that "future generations do not vote; they have no political or financial power; they cannot challenge our decisions".[109] Hence, it is ultimately important for the development of a legislative mechanism to represent future generations,

---

107 Academics and policy-makers claim that community-based organisations can mobilise citizens to take on more sustainable behaviours. This entails organisations such as schools, places of worship, social groups, clubs and other community groups playing a part in persuading individuals to reduce their impact on the environment and on other people. Gardner, & Stern, 2002; Jackson, 2005.

108 The reasons to give voice to the unborn and to future generations are explained by Shelton: "A depleted environment harms not only present generations, but future generations of humanity as well. First, an extinct species and whatever benefits it would have brought to the environment are lost forever. Second, economic, social, and cultural rights cannot be enjoyed in a world where resources are inadequate due to the waste of irresponsible prior generations. Third, the very survival of future generations may be jeopardised by sufficiently serious environmental problems". Shelton, 1991, p. 110.
Several authors recognise the possibility to grant the voice to future generations, see in general: Beckman, 1994; Dobson, 1996; Elgar, Doeleman, & Sandler, 1998, p. 1; Skagen, 2005, p. 429; Epstein, 1988, p. 67; Estlund, 2003, p. 31; Ford, 1998, p. 142; Gardner, 1978, p. 9. H. Barry, 2003b, p. 31; Low, & Gleeson, 1998; Mahoney, 2002, p. 88; Bradford, 1996, p. 5; Reiman, 2007, p. 35; Solum, 2001, p. 35; B. Thompson, 2004, p. 44; D. Thompson, p. 12; P. Wood, 2000b; P. Wood, 2000a, p. 411.

109 Our common future, 1987 see also Beckman, 2007.

especially since the decisions that the individual and government make today will determine the initial welfare of future generations (Weiss, 1984, p. 272).

Some authors, and also the World Commission on Environment and Development, suggest implementing this option by setting up an *ombudsman* for future generations (Weiss, 1992, p. 25; L. Westra, 2006). This institution could take a step towards ensuring that the interests of future generations are considered either "by giving standing to a representative of future generations in judicial or administrative proceedings or by appointing and publicly financing an office charged with ensuring that positive laws conserving our resources are observed, with investigating complaints of abuse, and with providing warnings of pending problems" (Weiss, 1984, p. 272). Moreover, as Weiss has suggested "a representative of future generations should be granted standing to intervene in proceedings before international tribunals such as the international Court of justice, regional tribunals, national courts and administrative bodies, and state or provincial courts" (Weiss, 1984, p. 272).

2.3.3 Implementation through Representation of Nature

As seen above, a criticism of the human rights approach is that it is mainly anthropocentric and it does not take into account the inherent worth of nature, so the question is whether non-human creatures can be adequately protected by legislation without having previously been granted those rights (Waite, 2007, p. 395).

Attempts to overcome the anthropocentric approach are plentiful, among these, the concept of nature's rights has been well documented since its rise to prominence in 1972, following the publication of Christopher Stone's article "Should trees have Standing?" (Stone, 1972, p. 450). For almost 40 years the concept has been debated amongst lawyers, philosophers, theologians and sociologists. This debate has led to an advocacy of a wide variety of rights approaches including legally enforceable rights for nature as envisaged by Stone. The point they have in common is an attempt to give concrete and meaningful recognition to the intrinsic value of nature.

Stone himself recognises the limitations of his 'rights' theory and in the final pages of his article discusses the importance of a changed environmental consciousness. He states that legal reform, together with attendant social reform, will be insufficient without a "radical shift in our feeling about 'our' place in the rest of Nature". Stone has never considered 'rights' as an end in themselves but rather as a means to an end.

There are many cases of laws whose immediate aim is the protection of the environment; but the final goal is always the benefit of the human race: for instance, the International Convention for the Regulation of Whaling which has now become a genuine species conservation treaty, the Ramsar Convention

1971, the EC Birds Directives, the EC Habitats Directive and the Convention on Biological Diversity 1992.

In the aforementioned case, the attribution of rights to the natural environment is ultimately anthropocentric in that it views the goal of environmental protection from a human perspective. However, the importance of rights for the environment lies in the express acknowledgement that the environment has intrinsic value (Waite, 2007, p. 395; P. Taylor, 1998, p. 209; Turner, 2009, p. 50).

2.3.4 Implementation through Ecological Limitations

Another way to implement the ecological duties towards the earth is to introduce ecological limitations in the international human rights approach. The scope of these ecological limitations is to implement ecocentric ethics in a way that grants responsibilities and duties upon humankind and takes intrinsic values and the interests of the natural community into account when exercising human rights (P. Taylor, 2009, p. 100).

Each human right has some boundaries created to protect the rights of others and common interest. One is to prescribe a right together with duties, so that the limits of the right will be determined by the duties. Another boundary is to prescribe specific boundaries around specific rights.[110]

Sieghart (Sieghart, 1985, p. 80) states that the limitation must protect one or more of a restricted set of public interests such as national security, public safety, public order, public health, public morals and the rights and freedoms of others. It should give to these limitations a narrow interpretation. Moreover, there is a burden to demonstrate that the law is necessary and that it protects the specified interest or interests. These restrictions might be extended to include ecological limitations consistent with recognition of an ecocentric approach (P. Taylor, 2009, p. 101).

Indeed, this idea of ecological limitations goes beyond environmental protection for the sake of human interests. Ecological limitations could be implemented following the standard formulations for boundaries to rights and freedoms. Several possibilities exist, including: imposing "the right to the use and enjoyment of property, together with a duty not to cause harm to the

---

110 Article 29 of the Universal Declaration states for example "Everyone has duties to the community in which alone the free and full development of his personality is possible. In the exercise of his rights and freedoms, everyone shall be subject only to such limitations as are determined by law solely for the purpose of securing due recognition and respect for the rights and freedoms of others and of meeting the just requirements of morality, public order and the general welfare in a democratic society. These rights and freedoms may in no case be exercised contrary to the purposes and principles of the United Nations".

ecological integrity of the natural environment, prescribing the right to the use and enjoyment of property, together with responsibilities to protect and enhance the ecological integrity of the natural environment; prescribing the right to the use and enjoyment of property, subject to a specific, or general, limitation in the interests of the general welfare of both nature and humanity" (P. Taylor, 2009, p. 103). Such limitations could apply to a number of other human rights.

Moreover, reinterpreting or extending phrases such as "general welfare" (used in Article 29 (2) of the Universal Declaration) should include respect for ecological integrity; or reinterpreting or extending phrases such as "duties to the community" (used in Article 29(1) should also include duties to the natural and human communities.

Conclusion of Section II

In summary, this second section has examined the actors of environmental democracy and its essential characteristics: environmental and ecological citizens. Environmental citizenship is centrally defined by its attempt to recognise universal principles relating to environmental rights and it centrally incorporates these into law. Ecological citizenship, granting obligation upon the citizens, promotes fundamental incorporation of the interests of other species and future generations into processes of democratic consideration. This leads to challenges to extend the boundaries of existing political citizenship beyond "the formerly relatively homogeneous notions of the nation-state and national community that to date have determined formal citizenship" (Christoff, 1996, p.163).

CHAPTER II

Environmental Democracy in an International Context

The first Chapter of this book examined the theoretical construction of environmental democracy by underlining its form, space and actors. On the one hand, it was shown that the best form of democracy for achieving environmental goals would be a mixed fusion of deliberative and participatory democracy; on the other hand, on a spatial scale, environmental democracy should be constructed at a global and local level to better answer to global and local environmental problems.

With regards to actors, the important role of individuals has been acknowledged and, consequently, so has their ecological and environmental citizenship, with specific rights and duties, which should be integrated towards the construction of this new form of democracy.

Therefore, this Chapter will focus on this theoretical construction, in order to determine whether it exists totally or partially at the global level. At this level, the creation of environmental democracy would be achieved through international environmental law which encompasses treaties, the tools by which the international and global relationship is determined.

Thus, the situation and the steps already taken at the international level will be examined, in particular through the *Aarhus Convention on Access to Information, Public Participation in Decision-Making and Access to Justice in Environmental Matters.*[111] This treaty has been chosen to be analysed, since it could serve as a catalyst, if not as a model, for the democratisation of international environmental decision-making processes and for the construction of an environmental democracy at a global level (Marshall, 2006, p. 126; Redgwell, 2007, p. 163).

In the words of the former Secretary-General of the United Nations, Kofi Annan: " By far the most impressive elaboration of principle 10 of the Rio Declaration, which stresses the need for citizen's participation in environmental issues [...] As such it is the most ambitious venture in the area of environmental democracy so far undertaken under the auspices of the United Nations" (Annan, 2000).

---

111 See Convention on Access to Information, Public Participation in Decision- Making and Access to Justice in environmental Matters, Participants, June 25, 1998, 38 I.L.M. 517 (1999), entered into force Oct. 30, 2001.

Before analysing the substantive and procedural environmental rights and ecological duties in international law and in the Convention, the first section intends to outline the conceptual framework of "democracy" and "environment" in international law and in the Aarhus Convention.

The second section will deal with the three official pillars: access to information, participation and access to justice, which are recognised by the Aarhus Convention; and the two additional pillars: enforcement of environmental law and the review of compliance mechanism. The aim of this part is not to consider all of the detailed provisions of the Convention but rather to take a broader look at the Aarhus version of the theoretical model of the environmental democracy and, hence, to show how the Aarhus pillars represent concrete examples of tools which could help to introduce some elements of the theoretical model of environmental democracy at an international level.

Section I: Elements of Environmental Democracy at the Global Level

*"Il ne sera jamais trop tard pour tenter de bien faire,
tant qu'il y aura sur terre un arbre, une bête ou un homme"*
(Yourcenar, 1980).

1 "Democracy", "Environment" and "Actors" in International Law and in the Aarhus Convention

The first step in order to better understand environmental democracy in an international context is to explain how this level deals with the following three elements: democracy, environment and actors.

1.1 Democracy: An Environmental Governance

The environment has become an issue of international concern since environmental problems are inherently global in character and affect every nation on Earth. Consequently, the nation-state level is no longer necessarily the most appropriate level to solve these kinds of problems. The international community, hence, has tried to organise itself in the face of the degradation of the environment and its resulting environmental catastrophes by using International Environmental Law, which encompasses environmental conferences, summits, declarations and treaties. The outcome of the above-mentioned instruments has been the creation of a new form of governance called "environmental governance".[112]

The word "governance" is a term to describe decision-making processes that are less formal than a government (Charnovitz, 2003, p. 183; Weiss, 2000, p. 345), and environmental governance are these processes in the environmental field at a global level.

Since the Stockholm Conference,[113] this new form of governance has begun appearing, and it was quickly received with enthusiasm by international literature, as a new global strategy to solve the ecological crisis of the Planet. Unfortunately, quite soon the optimistic opinion about its effects began fading (Finger, 2008, pp. 38-39).

1.1.1 Failure of Environmental Governance

The current situation looks vastly different from the euphoric year of 1972 which represents the birth of the first Conference and Declaration on the Human Environment at Stockholm or from the epic summer of 1992, when the world's governments came together at the United Nations Conference on Environment and Development (UNCED), and produced an "official stance of cautiousness" with regard to the world's environmental problems (Park, Conca, & Finger, 2008, pp. 4-5). UNCED evoked, indeed, an optimistic moment when the governments "were starting to come together to hammer out a cooperative path toward long-term sustainability. In this highly stylised vision, governments of the North and South shared a basic interest in responding to a set of problems linking environment and development goals" (Park, Conca, & Finger, 2008, pp. 4-5).

The first visible sign of inadequacy of this model could be seen after the 1992 Earth Summit in Rio de Janeiro,[114] and in particular in 2002 where at the world summit on sustainable development in Johannesburg a serious environmental agenda was almost entirely missing. 2009 marks the year of

---

112   About environmental governance see in general: Barstow Magraw, & Hawke, 2007, p. 614; Bernstein, 2005, p. 139; Bodansky, 2007, p. 704; Dunoff, 2007, p. 85; Kingsbury, 2007, p. 63; Stone, 2007, p. 291.

113   Declaration of the U.N. Conference on the Human Environment, Stockholm, Sweden, 5-16 June 1972, U.N. Doc. A/CONF.48/14/Rev.1, §1.

114   Rio Declaration on Environment and Development, Rio de Janeiro, Brazil, 3-14 June 1992, U.N. Doc. A/CONF.151/26/Rev.1.

failure of the Copenhagen Conference which was intended to find a new series of commitments to prevent Climate Change.

Moreover, August 21st, 2010 has been branded as an unfortunate milestone: the day in which we exhausted our ecological budget for the year. By the end of this day, humanity had demanded all the ecological services – from filtering CO_2 to producing the raw materials for food – which nature could provide for this entire year. From that point until the end of the year, we met our ecological demand by liquidating resource stocks and accumulating carbon dioxide in the atmosphere.[115]

Surely, much has been achieved to ameliorate certain elements of international environmental governance in the past years and several reforms have been suggested by the international doctrine; nevertheless, there is no negating that the scope and pace of change have been a source of major delusion so far.

The failure of present approaches to global environmental governance raise two questions. First, how is it possible, despite all the mentioned conferences and summits on the Human Environment, that the great global challenge of protection of the planet and its peoples has reached a point of such political and social insignificance? Second, if the path from Stockholm to Rio to Johannesburg cannot provide the basis for a serious approach, what is the alternative? (Park, Conca, & Finger, 2008, pp. 4-5).

Concerning the first question, it has been argued that one of the main weaknesses of the environmental law regime along with environmental governance is that it is a purely inter-governmental system. In fact, international environmental law is nothing more, or less, than the application of international public law to environmental problems.

Current environmental law is a result of tumultuous negotiations between states where individuals do not constitute international actors on the environmental scene. The reason is that historically the Westphalian tradition of international law allowed only national governments to deal with international matters.[116]

Citizens had no direct role, and their interests could only be considered to the extent that their government espoused them. Nevertheless, relying exclusively

---

115 Available at www.footprintnetwork.org/en/index.php/GFN/page/earth_overshoot_day. Every year, Global Footprint Network calculates nature's supply in the form of biocapacity (the amount of resources the planet generates) and compares that to human demand: the amount it takes to produce all the living resources we consume and absorb our carbon dioxide emissions. Earth Overshoot Day, a concept devised by U.K.-based new economics foundation, marks the day when demand on ecological services begins to exceed the renewable supply.

116 Named after the Peace of Westphalia that ended the Thirty Years' War over three centuries ago.

on national governments has proved to be inadequate. This system is also poorly designed to manage the ecological crisis due to the fact that national governments are extremely reluctant to bring environmental claims against other nations. Thus, national governments rely exclusively on themselves for compliance. This international legal system ignores the potentially powerful role that citizens can and do play in environmental law and policy.[117]

The state-centric system[118] of global environmental governance impedes the democratic aspirations of such citizens who would participate in political life, in particular in the creation of new laws in the context of the international community (Anton, 2008). For this reason several authors have spoken about "democratic deficit" at the international level. Thus, the first question can be answered in that failure of environmental governance is based mainly on a lack of democratic public participation at the international level. A second reason could be also added, according to Bosselman: the anthropocentric character of international environmental law.

Concerning the problem of democratic deficit, it is clear that a full analysis of this topic would go beyond the scope of this book; nevertheless, it must be pointed out that any democratic content in international environmental law is always derivative rather than direct and participatory. International environmental agreements are ineffective in particular because they do not capture any normative *consensus* among the individuals whose behaviour those agreements seek to regulate.

Hence, the democratic deficit is troubling as a general matter, but it is especially problematic in the area of environmental decision making if one accepts the premise that democracy, and in particular a participatory democracy, is a constitutive element of environmental democracy (Baber, & Bartlett, 2009, p. 103).

---

117 The international joint commission, the leading national environmental governance institution in the United States-Canadian environmental law regime, has recognized the importance of citizens in advancing compliance "public support is crucial to restore and protect the environment. N. D. Hall, 2007, p. 131.

118 Najam, Papa, & N. Taiyab, 2006. See, e.g., International environmental governance, Issues Paper for Ministerial Dialogue, 12th Session of the African Ministerial Conference on the Environment, U.N. Doc. EP Ref: III/1 (20 May 2008).

1.1.2 Solutions to the Inadequacy of Environmental Governance

1.1.2.1 Public Participation at Environmental Governance

To answer the second question, it can be said that the alternative to this system is a democratisation of the international environmental governance, which means to progressively include non-governmental organisations (NGOs), civil society and individuals in the law-making process. This means that nation-states have to collaborate increasingly with non-state actors to achieve ecological goals (Bosselmann, 2008, p. 175).

The first Chapter has highlighted how active public involvement may have significant consequences for the environment (Hall, 2007, p. 131). Hence, rather than marginalising civil society's voice, the idea is to introduce more active forms of participation which could decrease the democratic deficit. Indeed, if it is the democratic deficit that robs international law of its moral authority and gives states the room they need to evade their international obligations, then the solution must be, at least in part, to provide a direct, participatory element for the world's citizens in the making of international law.

The first step to avoid democratic deficit is the recognition of the role of individuals (Baber, & Bartlett, 2009, p. 103). In 2000, the former UN Secretary-General Kofi Annan presided over the preparation of a report entitled, "We the Peoples: The role of the United Nations in the 21st Century", which declares, "The international public domain [...] must be opened up further to the participation of the many actors whose contributions are essential to managing the path of globalisation" (Baber, & Bartlett, 2009, p. 103). A suggestion on how to fulfil the above-mentioned objective is through deliberative and participatory approaches which develop regulatory mechanisms that might bring a more direct and participatory form of democracy.

Second, to be effective, international environmental law would have to encourage the formation of a transnational collective will in the absence of a sovereign authority; it would have to incorporate local ecological knowledge, and to respect a regulative norm of democratic *consensus*.

Thus, in terms of pushing the democratisation of the international system further, it should be moving to expand modalities of participation in the development of international environmental law to the level of the individuals. Taking small steps, participation ought to be conceptualised as including notification and comment rights for individuals in connection with the normative activities of intergovernmental organisations, as well as those associated with conferences and meetings of parties mandated by multilateral environmental agreements (Anton, 2008, p. 8).

International doctrine in general works against recognition of possible public participation in decision-making because it holds to a positivist view of international law and challenges the legitimacy of non-state participation in a process seen as being reserved solely for state actors.[119] Against this background, non-state actors have made continuing claims for access to international law making and institutions, and these demands have slowly but increasingly been met (Dannenmaier, 2007, p. 38). Some steps towards more participation have been made.[120]

In fact, the idea of individual participation in international environmental law-making is not a novelty and it is premised on the statement in Principle 10 of the Rio Declaration *"environmental issues are best handled with the participation of all concerned citizens, at the relevant level"* and in its implementation by Aarhus Convention.

Although with increasing frequency non-state actors are finding voice and influence within meetings of Treaty Conferences of Parties and International Commissions[121] and even some international organisations have started to involve them, the practice of engaging non-state actors is not sufficiently widespread and does not flow from a sense of obligation, or *opinio juris*, and thus cannot be seen as a part of customary law and consequently, in general, individuals are currently still excluded.[122]

---

119   See for instance: Aderson, 2005, p. 1255; Bolton, 2000, p. 205; Dannenmaier, 1997, p. 111.

120   Explanatory is the catalogue written by Charnovitz on one hundred years of growth in non-state participation, which demonstrates the variety and depth of access. Charnovitz, 2006a, p. 348.

121   Indeed, James Rosenau observes that even though the decisions in the international system „may be formally adopted by the votes of states, [...] their substance is in many ways a product of pressures from outside to which they have been subjected by diverse non-governmental constituencies". The flip side of this observation is that it is precisely because of the lack of decision-making authority that non-state actors have to pressure states to do what they cannot.

122   FOOD AND AGRICULTURAL ORGANISATION, THE STRATEGIC FRAMEWORK FOR FAO 2000-2015, para. 118 (1999). "The FAO is an increasingly important Organisation in the field of international environmental law. It is significantly involved with fisheries, forests, biodiversity (including plant genetic resources), access to land and natural resources, and the worrying problem of food security. The FAO Strategic Framework 2000-2015 recognises the importance of public participation in its activities. In particular, the Strategic Framework commits the FAO to engage Civil Society Organisations (CSOs) and Nongovernmental Organisations (NGOs) in "policy dialogue at the country, regional and global levels, including improved access to FAO technical meetings". In practice, participatory openings and processes are fluid and vary across and within the FAO and its departments and divisions, but they remain focused on "stakeholders" – CSOs, NGOs and private sector (business) Organisations – rather than individuals". Anton, 2008, p. 8.

1.1.2.2 Aarhus Convention Model

The Aarhus Convention, fully titled "The United Nations Economic Commission for Europe Convention on Access to Information, Public Participation in Decision-Making and Access to Justice in Environmental Matters" was signed on 25 June 1998 in Aarhus, Denmark, and entered into force on 30 October 2001, ninety days after the deposit of the sixteenth instrument of ratification, and as of 26 September 2012, it counts 46 Parties.[123]

Article 1 states: "*In order to contribute to the protection of the right of every person of present and future generations to live in an environment adequate to his or her health and well-being, each Party shall guarantee the rights of access to information, public participation in decision-making, and access to justice in environmental matters in accordance with the provisions of this Convention*".

This provision underlines, unlike most multilateral environmental agreements which cover obligations that Parties have to each other, that the Aarhus Convention imposes a clear obligation on its Parties and public authorities towards the "public", as far as access to information, public participation and access to justice are concerned (Stec, & Casey-Lefkowitz, 2000, p. 1).

Article 1 also outlines the role of the State in reaching this goal. Thus, it is up to the State to provide for the necessary administrative, legal and practical structures, which shall guarantee the basic three rights, covered by the Convention. This represents a new approach to the role of the State. Instead of solving ecological problems itself, the State acts as a sort of referee in a process involving larger social forces, leading to a more organic and complete result.

1.1.2.2.1 Notion of "Democracy" in the Convention
This Convention was conceived with the express aim of promoting democracy to establish this kind of governance and to solve, at the same time, the problem of democratic deficit, as well as protecting the rights of everyone to live in a healthy environment. The agenda of Aarhus was mainly determined by a concern to get the newly independent states of Central and Eastern Europe closer to the European Union standard of environmental protection, and the "democratic component was considered crucial to this end" (Hayward, 2000, p. 144).

Consequently, drafters and commentators alike claimed that "the experience of the former communist bloc testified to a direct correlation between deficits in democracy and environmental quality. In fact, access to reliable information on

---

123   The signatory nations include almost all the nations of Europe and most of the former Soviet Socialist Republics, but not Russia. For a list of nations participating in the treaty, see Convention on Access to Information, Public Participation in Decision- Making and Access to Justice in environmental Matters, Participants.

the environment and recognition of the role of NGOs in raising the level of public awareness of environmental issues were seen as prerequisites to developing a civil society of democratic citizenry" (Hayward, 2000, p. 144).

For this reason, in its preamble, the contracting parties state their belief "*that the implementation of this Convention will contribute to strengthening democracy in the region of the United Nations Economic Commission for Europe*". Moreover, the ministerial declaration of the Aarhus Conference, by which the Convention was adopted, praised the Treaty as a significant step forward both for democracy and for the environment.[124]

In addition, there are two paragraphs in the Convention's preamble which confirm this will. Paragraph 8 expresses and calls attention to the "protective role of democracy" (Bell, 2004, p. 101). The idea of this paragraph is that the public needs participation rights to protect itself from the institutions of representative democracy, whereas paragraph 9 established that the institutions of representative democracy need public participation in order to achieve better environmental outcomes (Bell, 2004, p. 101).

While it is clear by the terms of the Convention that democracy is fundamental for a government to reach environmental goals, it has pointed out that Aarhus' notion of democracy is based "on the right to information as a prerequisite for the effective exercise of the rights to political participation and access to justice, which complete the "triptych" of the procedural rights of "environmental democracy". In a sense, these rights "give practical form, in the specific context of environmental policy, to the general principles of democracy and the rule of law already enshrined in other international prescriptive instruments on the protection of human rights" (Pallemaerts, 2003a).

1.1.2.2.2 Form of Environmental Democracy in the Convention
It has been said in the first Chapter that, fundamentally, from a theoretical viewpoint, it is the representative democracy with elements of participation which can better answer to the ecological crisis to construct an environmental democracy. The Aarhus model of democracy offers a concrete example (Bell, 2004b, p. 101).

To recall the main characteristics of deliberative and participatory theories, they both seek to improve the substantive output of regulation efforts as well as to introduce a plurality of perspectives into the decision-making process.

---

124 Declaration by the Environment Ministers of the region of the United Nations Economic Commission for Europe (UN/ECE), 4th Ministerial Conference „Environment for Europe", Aarhus, Denmark, 23-25 June 1998, para. 40. Pallemaerts, 2004, p. 19.

On the one hand, participation might improve the quality of decisions by way of enabling input from a wide range of participants that influence environmental policy making.   On the other hand, the ideal of deliberation rests broadly on the notion that through the rational debate of citizens, arguments are refined and preferences are transformed, leading both to improved solutions and real democratic engagement decisions. The deliberative approach is supposed to absorb the most selfish instincts of the individual or groups. (Lee, & Abbot, 2003, p. 83).

Moreover, the most important common point is that they both grant the right to citizens to participate, and they consider that general public involvement will produce a better answer to the environmental crisis, one which has been carefully reflected upon (Lee, & Abbot, 2003, p. 83).

When analysing the Aarhus Convention from this perspective, wide participation of the public mainly through participative mechanisms, but also through some deliberative tools, can be recognised. The analysis of the Convention in the second Section of this Chapter will clarify it further.[125]

Although the introduction of participatory rights brings a fundamental element of participatory democracy to the environmental decision-making process, the Aarhus ideal is to supplement representative democracy with participatory democracy (Bell, 2004b, p. 98).

The supremacy of representative institutions and the narrow role of public participation are evident throughout the Convention. The participation process is convoked and controlled by predetermined public authorities. The role of the public appears limited to a contribution to the technical assessment of the alternatives, and the public authorities determine which of the alternatives are the best "means" to a pre-defined "end". Thus, the institutions of representative democracy are the authoritative decision makers.

The public is not given the right to initiate a participation process, organise its format or set its timetable. According to Article 6(8), public authorities must follow the consultation procedure and they must "ensure that in their decision

---

125   The twentieth paragraph states that "Convinced that the implementation of this Convention will contribute to strengthening democracy in the region of the United Nations Economic Commission for Europe (ECE)". The links between the Convention and democratisation are made clear by the Chairman's Summary of the Seventh Economic Forum of the Organisation for Security and Cooperation in Europe (OSCE) (May 1999). That document urged countries to ratify the Aarhus Convention to affirm their commitment to public participation. The meeting considered that the matters at the heart of the Aarhus Convention were important for security in Europe, and recommended that the principles of the Aarhus Convention should be incorporated into an OSCE charter on European security. The report of the relevant working group was accepted even by States that had not signed the Aarhus Convention", Stec, & Casey-Lefkowitz, 2000, p. 23.

due account is taken of the outcome of the public participation". "However, they are not bound to accept or act on the comments of any of the participants".

It can be said that taking into account the outcome of public participation does not require the relevant authority to accept "the substance of all comments received and to change the decision according to every comment" (Stec, & Casey-Lefkowitz, 2000, p. 109). The public authorities should treat comments from the public as "information to be added to the information that they have from other sources" (Stec, & Casey-Lefkowitz, 2000, p. 109).

The wide liberty allowed to the institutions of representative democracy supports the interpretation that the "democratisation" stipulated by Aarhus is not a radical attack on representative democracy. At most, it is an attempt to face the democratic deficit with respect to concrete environmental matters and a small step forward towards opening up the "elite dominated policy-making processes" (Bell, 2004b, p. 98).

Hence, from a formal viewpoint, despite limiting them into premised boundaries, the Aarhus Convention recognises elements of environmental democracy, among which certain mechanisms deriving from the participative and deliberative model of democracy.

1.1.2.2.3 Space Scale of Environmental Democracy in the Convention

From the spatial dimension point of view, the Convention aims at influencing international practice beyond the limits of the United Nations Economic Commission for Europe (UNECE). Despite this treaty being principally a vehicle for promoting public access at a domestic level, the spatial approach of the Convention has also been reinforced by global ambitions. In the words of the former Secretary-General: "Although regional in scope, the significance of the Aarhus Convention is global".[126]

There is a wide selection of evidence for this approach. First, proponents of the Aarhus Convention have often looked at the possibility of exporting the agreement. The implementation guide notes that the convention is "open to accession by non-ECE countries, giving it the potential to serve as a global framework for strengthening citizens' environmental rights".

The global relevance of the Convention is further confirmed by Article 19(3): "Any other State, not referred to in paragraph 2 above, that is a Member of the United Nations may accede to the Convention upon approval by the Meeting of the Parties". According to this provision, membership is not only open to State

---

126   This actually exemplifies the 'desire of UNECE to continue to produce environmental agreements that are at least potentially beneficial at the global level'. See also Morgera, 2005, p. 138.

members of, or States having consultative status with the UNECE but also to any other State which is a member of the UN, upon approval by the Meeting of the Parties.

In addition to the extra-regional aspirations of its proponents, the Convention itself contains a unique provision that appears to be a call to a more international scale: Article 3(7) states "Each Party shall promote the application of the principles of this Convention in international environmental decision-making processes and within the framework of international organisations in matters relating to the environment".

This Article explicitly requires State parties to promote the application of the principles of the Convention in "international environmental decision-making processes" as well as in matters relating to the environment within the framework of international organisations (Morgera, 2005, p. 138). International environmental decision-making processes include bilateral or multilateral decision-making related to shared natural resources, as well as the decisions of bodies established through International Conventions.

According to the Implementation Guide of the Aarhus Convention it should also include "conferences of States on environmental issues, such as the 1992 Rio Conference or the periodic ministerial meetings "Environment for Europe" or "Environment and Health". Working groups charged with the negotiation of international legal instruments would also fall under this category".[127]

Thus, in matters relating to the environment, Parties are also obliged to promote the Aarhus Convention's principles in respect of international organisations. This norm obliges Aarhus parties to begin to handle the democratic deficit associated with the negotiation of international treaties and, especially, the operations of international governmental organisations, which are, as Bell has noted "'infamous' for their secrecy and their remoteness from the public" (Bell, 2004b, p. 100).

Such organisations include, according to the Implementation Guide "multilateral lending institutions such as the European Bank for Reconstruction and Development, specialised agencies and other organisations in the United Nations system such as the World Bank and the World Trade Organisation,

---

127 The drafting of the Protocol on Water and Health to the Convention on the Protection and Use of Transboundary Watercourses and International Lakes is one process in which many of the principles of the Aarhus Convention have already been applied. The Protocol's negotiating parties expressly took the Aarhus Convention into account. This may be contrasted with the Protocol to the Convention on Long-range Transboundary Air Pollution to Abate Acidification, Eutrophication and Ground-level Ozone, which has good active information provisions but did not follow Aarhus principles on passive information. Stec, & Casey-Lefkowitz, 2000, p. 45-47.

and special international organisations formed for specific tasks, such as the reconstruction of post-war infrastructure in the Balkans".[128]

The commitment laid out by Article 3(7) is a unique feature of an equally unique international accord. It suggests on the one hand that citizens can expect from their governments to advance the principle of participatory democracy on the international stage and on the other hand, the citizens from non-state parties might benefit from Aarhus principles when engaging in international forums on environmental issues (Dannenmaier, 2007, p. 33).

Despite the fact that the form in which such obligation might be transferred or inherited would remain uncertain, it is clear that no commitments would be assumed by international institutions automatically; nevertheless the expression "obligation to promote" has been seen as an obligation that might be characterised "as a duty of evangelism" (Dannenmaier, 2007, p. 47). The meaning of this term shall not be interpreted in a religious way; but it underlines the Aarhus parties' commitment to promote environmental democracy, which they have embraced domestically and internationally.

In addition, in 2005, the Aarhus contracting parties developed the Almaty Guidelines[129] which are designed to provide further guidance on the promotion of the application of Aarhus in international *fora*. The Guidelines are vaguely drafted and have an exhortatory character.[130] Consequently, there is no real

---

128 Stec, & Casey-Lefkowitz, 2000, p. 45-47. As Pallemaerts has recalled (Pallemaerts, 2004, p. 20): "In the Lucca Declaration adopted at the first meeting of the parties to the Aarhus Convention, ministers "recognise[d] the need for guidance to the Parties on promoting the application of the principles of the Convention in international environmental decision-making processes and within the framework of international organisations in matters relating to the environment and [...] therefore recommend[ed] that consideration be given to the possibility of developing guidelines on this topic". Acting pursuant to this ministerial mandate, the Working Group of the Parties to the Convention, at its first meeting in November 2003, decided to establish an *ad hoc* expert group "to consider the scope, format and content of possible guidelines and the appropriate process for their development". (Report of the first meeting of the Working Group of the Parties, UN Doc. MP.PP/WG.1/2003/2, 26 November 2003, p. 9, para. 47). This group had its first meeting in June 2004 and in the second meeting of the parties, in May 2005 in Almaty, Kazakhstan, was adopt a set of Guidelines was adopted concerning the implementation of article 3.7: "these Guidelines are intended, through their application, to positively influence the way in which international access is secured in international forums in which parties to the Convention participate" (para 6) and then, the Guidelines may "also serve as a source of inspiration to Signatories and other interested States" (para 3).

129 Report of the 2nd meeting of the parties, ECE/MP.PP/2005/2/Add.5, Decision II/4 of 20 June 2005 promoting the Application of the Principles of the Aarhus Convention in international forums.

130 Report of the 2nd meeting of the parties, ECE/MP.PP/2005/2/Add.5, Decision II/4 of 20 June 2005 promoting the Application of the Principles of the Aarhus Convention in international forums.

obligation imposed to the parties, and the very short section on review procedures in environmental matters "provides that each Party should encourage the consideration in international forums of measures to facilitate public access to review procedures relating to any application of the rules and standards of each forum regarding access to information and public participation within the scope of these guidelines" (Schall, 2008, p. 417).

The inclusion of Article 19 and Article 3(7) shows the wide spatial scale of the Aarhus ambitions to democratise all levels of governance, global and local. In addition, the last provision, Article 3(7), emphasises also a movement towards an introduction of a new form of citizenship at the international level.

Article 3(7), thus, not only places parties to the Aarhus Convention in a position that "favours a more meaningful democratic process at the international level, but it would also appear to adopt a rather proactive stance in the promotion of this democracy – and in the construction of some form of global environmental citizenship" and ecological citizenship.[131]

Hence, citizens can, under the Aarhus Convention, claim a "degree of citizenship at the international level – speaking with their own government and with other states about environmental concerns and interests (Dannenmaier, 2007, p. 49).

1.2 Notion of "Environment" at Global Level

It has been emphasised in the first Chapter that the relationship between man and nature is primarily anthropocentric and consequently also environmental law has taken this dominant approach: in other words, what is good for humans is good for nature.

In addition, concerning the reasons of the failure of environmental governance, some commentators have pointed a finger against the anthropocentric approach of international environmental law which does not acknowledge the fundamental

---

131   This could include, where a process is entirely governed by those states party to Aarhus, making formal changes to the organic commitments of the institutions or creating access protocols to establish more open and responsive procedures. It could also include where parties to the Aarhus Convention are a majority of states to a convention or institution, using this majority status to move the forum in a more democratic direction. Where only a few Aarhus parties are participants, they could still make an effort to adopt informal mechanisms [...] even as they promote more formal mechanisms for consideration by the broader forum. Finally, it could include a commitment by Aarhus parties to adapt their foreign policy in matters relating to the environment [...] to assure that the state's own delegations are open and transparent and that the positions taken and votes made by the state are consistent with the principles of the Aarhus Convention. Dannenmaier, 2007, p 50.

importance of the preservation of Earth's ecological integrity according to an ecocentric vision.

In order to reach this some authors suggest to transform environmental governance into governance for sustainability. Governance for sustainability is value-based, acknowledging the fundamental importance of the preservation of Earth's ecological integrity (Bosselmann, 2008, p. 175).

1.2.1 "Environment" in International Environmental Law: an Anthropocentric Approach

In fact, the emerging field of environmental law is being built on the basic platform of the service of human need and nature, ecosystems, natural resources, wildlife and climate change, which are areas of concern to international law-makers primarily for their value to humanity (Birnie, & Boyle, 2002, p. 5).

There are many examples of this view in international environmental law: the 1972 Stockholm Declaration on the Human Environment reflected its anthropocentric basis in its title. This was strengthened further with the emphasis upon protecting the environment for present and future (human) generations. Additionally, the Declaration emphasised that "of all things in the world, people are the most precious",[132] and it adds "both aspects of man's environment, that natural and the man-made, are essential for his well-being and enjoyment of basic human right" (Caldwell, 1980, p. 170).

The anthropocentric and instrumental view of nature can also be seen in the 1980 World Conservation Strategy, which affirmed "living and non-living resources" serve "to satisfy human needs and to improve the quality of human life".[133] According to this approach, "other species do not have intrinsic value in their own right, but are considered resources for human use" (Merchant, 1992, p. 229-230).

Likewise, the 1992 Rio Declaration on Environment and Development did not take a step back from this dominant position and stated "human beings are at the centre of concerns for sustainable development".

In a number of later international environmental documents, anthropocentric terminology appears designing nature as resources rather than attributing to it intrinsic value of its own accord.

Once it is stated that humanity is at the centre of environmental policy, and that everything non-human is to be regarded as resources, then all justification for environmental protection must come from anthropocentric considerations.

---

132 Para.7, Ch. I of the Declaration.
133 See for more information Rusta, & Simma, 1980, p. 427.

Also at the international level, nature will only be conceived "on account of the instrumental values attributed to it by humans, rather than being protected because of its own intrinsic value" (Gillespie, 1997).

Nevertheless, it would appear that new ethical bases of international environmental law and policy are arising but this approach is very marginal and it is still at a rudimentary stage from a legally binding point of view. Nowadays, indeed, only the recognition of the need for a new approach is evolving (Gillespie, 1997). Consequently, the content and implications of this new approach are gravely lacking except in few examples in international law, such as in the Earth Charter.[134]

1.2.1.1 Wide and Narrow Definition of Term "Environment"

Concerning the elements which compose the environment it is hard to find a unique and universally recognised definition, since international environmental law is generally devoted to disciplining specific subjects.[135] Authoritative scholars have remarked that, from a legal point of view, a comprehensive idea of the environment does not exist and environment rests a "fuzzy concept".[136]

---

134 The Draft Earth Charter is published in the Pacific Institute of Resource Management, 1992, Commitment for the Future: The Earth Charter and Treaties agreed to by the International NGOs and Social Movements, Paper Presented to the International NGO and Social Movements Forum Conference, Wellington, New Zealand, 11 June.

135 An example of a definition is Article 7(d) of the Convention on Long-range Transboundary Air Pollution of 1979 (Convention on Long-range Transboundary Air Pollution of 13 November 1979, 18 ILM 1979, p. 1442). According to this provision the environment includes agriculture, forestry, materials, aquatic and other natural ecosystems and visibility. This definition is a good example for the statement made initially that the definitions given will serve the context they are made in: Thornton, & Beckwith, 2004, p. 5.

136 Birnie, Boyle, & Redgwell, 2009, p. 603. Given the absence of an official and authoritative definition of environment at the international as well as European level, the doctrine has suggested different categories. Kiss has stated that in law "environment" can refer to a limited area or encompass the entire planet, including the atmosphere and stratosphere (Kiss, & Shelton, 2000b). In comparison, Rodgers uses the categories of "human", containing health, social and other man-made conditions, versus "natural" environment, including the physical condition of the land, air and water (Rodgers, 1977). Three groups have been suggested by Salter. Firstly, under the heading of "natural" environment, the protection of environmental media is captured; the second category is the "man made" environment including cultural heritage. And finally, a third category concerns "human" environment, including regulation on food, products, safety issues, leisure and economic health as consumer protection, eco-labeling, and so forth (Salter, 1994).

Cano also has classified the total environment as being composed of three categories: the natural environment made up of natural resources; the created environment, e.g., things or institutions created by mankind; and the induced environment, e.g., the results of man's utilization of natural resources in agriculture, forestry, animal husbandry, pisci-culture, etc. (Cano, 1975).

In Caldwell words, environment "is a term that everyone understands and no one is bale to define" (Caldwell, 1980, p. 170). Since all human activities have an effect on the environment, it is difficult to reach an all-inclusive final definition of this concept.[137]

Despite the vagueness of this term, it is possible to find two categories of definitions: one includes only natural elements, a narrow definition, and the other includes a social dimension, a wide definition. The following shall give an overview over that. First, however, the term natural resources shall be addressed particularly since they constitute part of both sets of definitions.

The definitions of natural resources at an international level can be classified in two sets (Reiners, 2009). It has been suggested that natural resources are naturally providing materials that are useful to man (Skinner, 1986, p. 1). Another proposal is that natural resources are tangible or intangible goods that may be used in an economic manner or to create economic value and that are not manufactured or produced (Rosenne, 1986, p. 64). These definitions imply

---

Sands has underlined four possible elements which can be found in international acts: "fauna, flora, soil, water and climatic factors; material assets, including archaeological and cultural heritage; the landscape and environmental amenity; and the interrelationship between the above factors" (Sands, 1995, p. 629).
Jayanti, 2009, p. 1. He distinguishes the Earth and the Environment: "Earth- the non-human and human created contents of the planet, including the planet itself" and " Environment – the planet-wide ecosystem that consists of the inter functioning of all the earth's ecosystems and natural elements, including, but not limited to, the oceans, forests, atmosphere, mineral deposits, fresh bodies of water and waterways, coastal regions, deserts, tundras, fisheries, mammal populations, bird flocks and other flora, fauna and natural resources".

137   By environment, an object regulated by the environmental law, it means in some cases "only the natural environment which is composed by the natural goods, the water, the sea, the air, the flora and the fauna, and in general whatever surrounds man and has been created without this agency, while in other cases there is a visible and marked generalisation of the term's application which result in the conclusion that everything is environment. In these latter cases, the legal protection of the environment also embraces the built environment which is subcategorised into urban and cultural and comprises man-made structures". "Public law is considered to be the most suitable means for the effective protection of the environment, because it constitutes the main means for the enforcement of the public interest, part of which is the protection of the environment, one of the most important in width and in essence areas of modern state intervention. The law may include the conditions and the procedures for the prevention and restoration of environmental damages and for the restriction of the full ban of polluting activities. The role of Private law should not be underestimated. Through it every person has the right to act individually in order to protect his living space, in order to prevent environmental damage, taking advantage in order to achieve "the restitution of the damages of the environment reflecting on his individual goods on his life, his health and, generally, on his personality, and also to intervene inviting judicial control of the damaging activities and their originators, thus counterbalancing the frequent lack of interest and the complacent short-sightedness of public authorities". Karakostras, 2008, p. 1.

that the appearance in nature must have an economic value. This economically-valuing definition represents the first set of definitions.

The second category covers definitions of natural resources which do not include an economic element.[138] It was suggested that natural resources are all physical natural goods, as opposed to those made by man (Cano, 1975, p. 1). Hence, there are basically two sets of definitions, one requiring an economic value, the other excluding it. Nevertheless, both definitions agree on the fact that a natural resource is something nature-given, so to speak, and not man-made. Moreover, they do not seem to include human beings.

For example, Principle 2 of the 1972 Stockholm Declaration states "The natural resources of the earth, including the air, water, land, flora and fauna and especially representative samples of natural ecosystems, must be safeguarded for the benefit of present and future generations through careful planning or management, as appropriate".[139]

Concerning the term "environment" the Dictionary of Environmental Law defines it in a narrow way as "The physical surroundings or circumstances in which humanity struggles to survive and thrive; it includes the planet Earth and outer space as well as the immediate province of living organisms, the biosphere. The environment of the individual includes the biotic factors of land, water, atmosphere, climate, sound, odours and tastes; and the biotic factors of other humans, animals, plants, bacteria and viruses" (Gilpin, 2000, p. 92- 93).

Despite the narrow definition, environmental legislation generally defines the term broadly, going far beyond the limits of the natural environment and permitting an assessment of almost any aspect of a proposed undertaking, including its impact on people and society as whole.

Hence, there is another major class of things in addition to the physical universe: abstract entities such as society. The two classes are amenable to further differentiation into living, organisms dependent upon the non-living, and non-living forms such as land, air and water.

The Stockholm declaration includes also a wide approach: "both aspects of man's environment, that natural and the man-made, are essential for his well-being and enjoyment of basic human rights". Thus, the man-made environment, including buildings, monuments or other structures and landscapes are considered as part of the environment to be protected (Thornton, & Beckwit, 2004, p.1).

---

138   This approach is favored by Schrijver, 1995, p. 15-16.
139   Stockholm Conference, declaration on the Human Environment, Principle 1, Report of the United Nations Conference on the Human Environment, New York, 1973.

Another example of a wide definition is the Convention on Civil Liability for Environmental Damage resulting from activities dangerous to the environment: "for the purpose of this Convention [...] Environment includes: a) natural resources both a biotic and biotic, such air, water, soil, fauna and flora and the interaction between the same factors; b) property which forms part of the cultural heritage; and the characteristic aspects of the landscape".[140]

Finally, the International Court of Justice includes a social dimension in the definition of the Environment, stating, "the environment is not an abstraction but represents the living space, the quality of life and the very health of human beings, including generations unborn".[141]

1.2.2 Aarhus Definition of the Term Environment

While the Convention does not attempt to define the terms environment or environmental matters, some suggestions of the significance of these expressions in accordance with the Convention can be derived from the contained definition of environmental information.

Article 2(3) lists what environmental information can encompass.[142] There are three categories: environmental information includes any information in material form relating to the state of the elements of the environment. The provision lists examples to illustrate what is meant by elements of the environment.

In this limited description the following elements are included: air and atmosphere, water, soil, land, landscape and natural sites, and biological diversity

---

140 Art. 2 of Convention on Civil Liability for Damage Resulting from Activities Dangerous to the Environment, arts. 13-16, June 21, 1993, 32 I.L.M. 1228.
Furthermore, some international treaties contain some form of a working concept for the environment by adopting definitions appropriate for their purpose. The 1979 Long-range Transboundary Air Pollution Convention, for example, has a definition of *"environment that includes agriculture, forestry, materials, aquatic and other natural ecosystems and visibility"*.

141 International Court of Justice, Advisory opinion of 8 July 1996.

142 Article 2 provides the following definition: "environmental information" means any information in written, visual, aural, electronic or any other material form on: (a) The state of elements of the environment, such as air and atmosphere, water, soil, land, landscape and natural sites, biological diversity and its components, including genetically modified organisms, and the interaction among these elements; (b) Factors, such as substances, energy, noise and radiation, and activities or measures, including administrative measures, environmental agreements, policies, legislation, plans and programmes, affecting or likely to affect the elements of the environment within the scope of subparagraph (a) above, and cost-benefit and other economic analyses and assumptions used in environmental decision-making; (c) The state of human health and safety, conditions of human life, cultural sites and built structures, inasmuch as they are or may be affected by the state of the elements of the environment or, through these elements, by the factors, activities or measures referred to in subparagraph (b) above". Stec, & Casey-Lefkowitz, 2000, p. 35.

and its components, including genetically modified organisms. It is also useful to look at other legal sources, which may be relevant in understanding the scope of the above elements. Regarding air and atmosphere, the EU Council Directive 96/62/EC of 27 September 1996 on ambient air quality assessment and management defines "ambient air as outdoor air in the troposphere, excluding work places".[143]

Furthermore, soil, land, landscape and natural sites are grouped together in order to, according to the Implementation Guide of the Convention, "ensure a broad application and scope" (Stec, & Casey-Lefkowitz, 2000, p. 35). There are several reasons that landscape and natural site safeguards are central elements of protection. Some examples are "aesthetic appeal, protection of unique historical or cultural areas, or preservation of traditional uses of land".

Moreover, natural sites may refer, as has been suggested "to any objects of nature that are of specific value, including not only officially designated protected areas, but also, for example, a tree or park that is of local significance, having special natural, historic or cultural value"(Stec, & Casey-Lefkowitz, 2000,p. 35).

The wording "biological diversity and its components" can be understood better in the light of Article 2 of the Convention on Biological Diversity, which gives the following definition of biological diversity: "the variability among living organisms from all sources including, inter alia, terrestrial, marine and other aquatic ecosystems and the ecological complexes of which they are part; this includes diversity within species, between species and of ecosystems".

Thus, biodiversity encompasses ecosystem diversity, species diversity and genetic diversity. Moreover, substantial entities, which are identifiable as a specific ecosystem, as a dynamic complex of plant, animal and micro-organism communities and their non-living environment interacting as a functional unit,[144] are considered elements of biodiversity.[145]

The reference to genetically modified organisms can be defined by Directive 2001/18/EC on the deliberate release into the environment of genetically modified organisms. According to the Directive, a genetically modified organism is "an organism in which the genetic material has been altered in a way that does not occur naturally by mating and/or natural recombination".[146]

---

143   OJ L 296 p. 55, 1996/11/21, Article 2(1).
144   Convention on Biological Diversity, Article 2.
145   A Guide to the Convention on Biological Diversity, Switzerland, 1994, p. 16.
146   Directive 2001/18/EC of the European Parliament and of the Council of 12 March 2001 on the deliberate release into the environment of genetically modified organisms and repealing Council Directive 90/220/EEC.

Furthermore, the definition includes the phrase: "the interaction among these elements", this sentence recalls the Bosselmann definition of environment under the term "ecological integrity", namely, "the interactions between the various life forms – including human beings – we should be concerned with" (Bosselmann, 2008).

This expression mirrors the meaning reflected by Integrated Pollution Prevention and Control, which acknowledged that the relations between environmental elements are as significant as the elements themselves.[147]

The catalogue of the elements of the environment is not complete and others can exist. For example, according to the Implementation Guide (Stec, & Casey-Lefkowitz, 2000, p. 35) "radiation, while being mentioned in subparagraph (b) as a "factor", may also be considered as an element of the environment. Otherwise, the effect of radiation on human health would be covered by the definition only if it acted through an environmental medium".

Finally, environmental information includes a third category of the elements: "the state of human health and safety, conditions of human life, cultural sites and built structures" as things which may be included under environmental information. "Conditions of life," in conformity with the Implementation Guide, generally "may include quality of air and water, housing and workplace conditions, relative wealth, and various social conditions" (Stec, & Casey-Lefkowitz, 2000, p. 50).

Hence, the Convention takes note of the fact that the human environment, including human health and safety, cultural sites, and other aspects of the constructed environment, tends to be affected by the same activities, factors or measures, which have an effect on the natural environment (Stec, & Casey-Lefkowitz, 2000, p. 33).

However, it should be remarked that the above-mentioned third category includes a social dimension, for example, like that which has been added by the International Court of Justice in its already mentioned definition of Environment.

To sum up, the Treaty includes a wide definition and even though it is not explicitly possible to affirm the anthropocentric approach, it characterises the basis of the Convention itself.

---

[147] Council Directive 96/61/EC of 24 September 1996 Concerning Integrated Pollution Prevention and Control. OJ 1996, L 257/26.

1.3 Actors of the Environmental Democracy at Global Level

As seen in the first Chapter, the role of the citizens is fundamental to build up an environmental democracy. In this part, it will be studied whether and how an individual or a group of individuals act to define the environment at an international level. In order to do this, it is necessary to briefly explore the notion of "stakeholders", which includes also the civil society and NGOs.

1.3.1 Actors in International Environmental Law: Notion of "Stakeholders"

At an international level, the term "stakeholders" has been used to include all the parties taking part in the international institutions' deliberative and decision-making processes.[148] The main parties concerned in these processes are obviously the Member States of these institutions. But, although the States remain the principal stakeholders, the notion of stakeholders extends also beyond the States formally involved in the decision-making processes.[149]

Recently, indeed, the Cardoso report on the UN has upheld the move under way towards an express recognition of different "stakeholders" in global governance; in particular, as much in the environmental and sustainable development fields as in other sectors where decision-making processes have traditionally been dominated by national governments and their representatives.[150] Nevertheless, this recognition has not yet necessarily found concrete expressions in an

---

148 Report from the Panel of Eminent Persons on United Nations-Civil Society Relations, Doc. A/58/817, 11 June 2004, p. 13. Also called Cardoso Report. See also: The Diversity of Actors Within the UN System, Background paper for the Secretary-General's Panel of Eminent Persons on Civil Society and UN Relationships, available at www.un.org/reform/pdfs/categories.htm

149 While states may have a diminishing role in international affairs, as compared to times past, states remain the main actors in the international system. It is true that the impetus for reform and innovation in international environmental law and policy in large measure comes from non-state actors, but most, if not all, the important decisions are still made by states. Schachter, 1997, p. 26-28.

150 Report from the Panel of Eminent Persons on United Nations-Civil Society Relations, Doc. A/58/817, 11 June 2004, p. 13. Also called Cardoso Report. For example, the Panel of Eminent Persons chaired by former President of Brazil, Fernando Henrique Cardoso, tasked by UN Secretary-General Kofi Annan with a study of the UN's relations with civil society, proposed three major categories of stakeholders: Member States, the private sector and civil society. Pallemaerts, & Moreau, 2004, p. 15.

extensive reform of the intergovernmental organisations' institutional rules andpractices.

The Cardoso report includes in the definition of stakeholders the private sector as comprising firms, business federations, employee associations and industry lobby groups, and civil society. The last group is defined as a sphere of social life that is public, but not part of the state, and the private sphere, but exclude profit-making activities.

A wide range of individuals and organisations can be embodied in this category. The Cardoso report includes "citizens' associations to which its members have decided to belong which promote their interests, their ideas and their ideologies, mass organisations, trade unions, professional associations, social movements, indigenous people's organisations, religious and spiritual organisations, academic associations and public-benefit non-governmental organisations (NGOs)" (Pallemaerts, & Moreau, 2004, p. 15).

It may be remarked that the Agenda 21[151] programme adopted at the Rio Summit also expressly recognises different components of civil society as belonging to the "major groups" whose role should be "strengthened" with a view to the "effective implementation" of its stated sustainable development goals, and, "moving towards real social partnership in support of common efforts". Agenda 21 specifically identifies the following "major groups": women, children and youth, workers and their trade unions, farmers, the scientific and technological community, business and industry, local authorities, indigenous people and their communities, and non-governmental organisations (Pallemaerts, & Moreau, 2004, p. 16).

This last group, including both the definitions of "stakeholders" and "major group", is the most prominent actor in the field of environmental governance at an international level and for this reason it necessary to further analyse its role.[152]

---

151   Agenda 21 of the United Nations Division for Sustainable Development, adopted at the United Nations Conference on Environment and Development (UNCED), Rio de Janeiro, Brazil, 3-14 June 1992.

152   For a more detailed analysis, readers are invited to refer to the more comprehensive studies by other authors: Beigbeder, 1992; Breton-Le Goff, 2001; Déjeant-Pons, 1999, p. 58; De Schutter, 1996, p. 372; Gemmill, & Bamiele-Izu, 2002; Green, 2004; Jordan, 2000; Mori, 2004, p. 157; Nelson, 2002; Pallemaerts, 2003b, p. 275; Paul, 2000; Schechter, 2001, p. 184; Speeckaert, 1956, p. 39-40.

1.3.1.1 NGOs in the Environmental Field

Individuals may play a fundamental role, on the one hand, in terms of their personal behaviour in protecting the environment and exercising procedural rights; on the other hand, they may also act in association with others, and the NGOs are the most famous expression of this.

A *consensus* about the important role an NGO could play in environmental democracy has been given by the Ministers for the Environment of the region of the United Nations Economic Commission for Europe: "We recognise and support the crucial role played in society by environmental NGOs as an important channel for articulating the opinions of the environmentally concerned public. An engaged, critically aware public is essential to a healthy democracy. By helping to empower individual citizens and environmental NGOs is to play an active role in the environmental policy-making and raising awareness".[153]

Concerning the definition of NGOs, it is not easy to provide one since the complexity and broad diversity of this social phenomenon makes it hard to pinpoint exactly what the NGO notion covers.[154] In fact, their operations encompass all economic and social activities and they can be found working at local and regional levels, at national and even at global levels. Despite these differences, some authors and organisations have tried to define such organisations: they are groups of persons or of societies which are freely created by private initiatives, which represent and pursue a specific interest and which are not directly profit seeking.[155]

---

[153] Fourth Ministerial Conference, Environment for Europe, Aarhus, Denmark 23–25 June 1998 Declaration by the Environment Ministers of the region of the United Nations Economic Commission for Europe (UN/ECE), available at www.unece.org/env/efe/history%20of%20EfE/Aarhus.E.pdf.

[154] In 1950, the International Law Institute drafted a convention in which it defined "international associations" as "groups of persons or communities freely created by private initiative who exercise, with a non- profit aim, an international public-interest activity that transcends any exclusively national concern".. The definition given by the European Convention on the Recognition of the Legal Personality of International Non-Governmental Organisations, adopted on 24 April 1986 with an entry into force in January 1991, is more precise. This convention drafted by the Council of Europe stipulates that the associations, foundations and other private institutions must satisfy the following conditions to be referred to as "international non-governmental organisations": "have a non-profit-making aim of international utility; have been established by an instrument governed by the internal law of a State; carry on their activities with effect in at least two States; and have their statutory office in the territory of a State Party and the central management and control in the territory of that Party or of another Party". Pallemaerts, & Moreau, 2004.

[155] The terms "NGO" and "social movement" are sometimes used interchangeably, but "NGO," where the term is used simply to refer to any non-state non-commercial organisation, is quite undiscriminating; social movements, as informal networks of actors

The recognition of the special *status* of NGOs appears consistent with the concern that ordinary citizens may often be ill-equipped to participate effectively in a process that will also include powerful stakeholders, such as businesses, local government and other executive agencies (Bell, 2004, p. 100). In particular, at the international level citizens might not have the time, money, knowledge or inclination to become informed and effective participants committed to enforcing their rights, in particular when the environmental issues concern not just a single country.

Moreover, NGOs often play a warning role and serve as a link between the scientific community by identifying ecological problems with their causes, and public opinion and the governments. As a "medium" between the individual and the State, NGOs can inform and call to account the international legal process.

They may represent interests that states do not take up, push the agenda of states forward in respect to issues which they do take up, or simply make alternative views available to inform and to enrich the decision-making process. For instance, interests represented in the environmental field might include those of minorities: for instance, indigenous peoples (Anaya, 2004), environmental refugees (Bell, 2004), and future generations or from an ecocentric point of view, nature itself.[156]

Furthermore, non-State actors can play important and very supportive roles in each step of the process of developing, implementing, and monitoring international environmental policies within international environmental governance.[157]

---

linked by a shared identity and engaged in collective action, are rarer and more complex. They may include NGOs but they cannot be reduced to the organisations that constitute (part of) their networks. See for this topic: Perrez, 2008.

156 NGOs are not in general constrained by property rights based on sovereignty and as such they are free to focus on un-owned interest such as the global commons; but of course there is still the risk that especially procedural rights, may be used by affluent groups or "cosmetic environmentalists" to protect a privileged quality of life, which may impose further environmental costs upon the dispossessed or environmentally vulnerable communities, which are in turn denied access to justice by poverty or lack of institutional skills.

157 1) By collecting, analysing, and disseminating relevant information, drawing the attention to new and emerging issues that need international attention and by mobilising public opinion through information campaigns and broad outreach activities, they can influence the *agenda-setting of international environmental governance.* 2) They can inspire and shape the *development of international norms and policies* by providing expert advice to state-centred international negotiations, formulating views and expressing interests that might be ignored by the State actors, by mobilising public opinion at the national level to influence the position of the representatives, by lobbying and monitoring governmental delegations during negotiations. 3) They can contribute to the good *understanding of international norms and policies* by public information, engaging in interpretation of

Nevertheless, it has been already said that NGO involvement and in particular individual participation is insufficient, and that international institutions and processes are criticised for still suffering from a democratic deficit.[158]

A step further towards a growing participation of such actors has been achieved by the Aarhus Convention.

1.3.2 Actors in the Aarhus Convention

1.3.2.1 The role of the citizens and NGOs

The Convention mentions the methods in which individuals may organise themselves in order to participate. In addition, it establishes in Article 3(4) a duty for States to provide for the acknowledgment of, and support of, associations, organisations or groups promoting environmental protection, in their national legislation (Morgera, 2005, p. 140).

The Convention has expressly emphasised that individuals and NGOs (Wates, 2005b, p. 395) can be both an essential player and a partner in the formulation and implementation of environmental policies (Pallemaerts, 2003a). This view can be found in the thirteenth preamble paragraph of the Convention which recognises "the importance of the respective roles that individual citizens, non-governmental organisations and the private sector can play in environmental protection".

The relevance of the role of the individuals and NGOs has been highlighted, in particular by the acknowledgement of their environmental rights and ecological duties, and additionally by Article 15 with its recent implementation. This entitles individuals and NGOs to participate in the monitoring procedure of State compliance with their treaty obligations, through the Aarhus International Body.

---

international rules and norms, and by contributing to international adjudication by making amicus curiae (friends-of-the-court) submissions. 4) They can support the *implementation of international environmental policies* by advising State actors, supporting State implementation and by performing operational functions themselves. 5) And, they can support *compliance with commitments and policies* by monitoring State action, by drawing the public attention to implementation problems, sue institutions at the national level for non-action, and, in specific environmental regimes, by triggering compliance procedures.

158   In fact, more controversially, arguments have been made for NGOs to have a role in compliance and dispute settlement on the international level, particularly in relation to the enforcement of environmental obligations. Cameron, & Mackenzie, 1996, p. 137. See also: Perrez, 2008. See also Charnovitz, 1997, p. 183; Charnovitz, 2006b, p. 348.

Furthermore, concerning NGOs, the Convention has acknowledged their important role in consideration of their inclination to make dynamic exercise of the rights thereby created, acting in general on behalf of the wider public.

It is worth noting that during the negotiations of the Convention a variety of non-governmental human rights and environmental organisations were broadly involved in the negotiation and drafting process, being authorised to participate in the plenary Sessions of working groups and in virtually every drafting committee, and to intervene at the discretion of the Chair (Bruch, & Czebiniak, 2002, p. 1428).[159]

The Aarhus negotiation process for this reason was unique. NGOs attended all the negotiation meetings, intervened in the debates and proposed even treaty articles. And EU Member State officials described the role of the NGOs like this: "they negotiated as if they were another country, quite a big country and they had an enormous impact on the negotiations" (Delreux, 2009, p. 330). As a consequence, the text of the Convention provides for a significant role for NGOs.[160]

In addition, as emphasised by the Resolution of the Signatories, for NGOs to effectively participate, they should be empowered to be involved "in the preparation of instruments on environmental protection by intergovernmental organisations other than UNECE, and encouraged international organisations, including the regional commissions of the United Nations and bodies other than UNECE, to draw upon the Convention to develop appropriate arrangements relating to the subjects covered by it" (Stec, & Casey-Lefkowitz, 2000, p. 47).

Another significant example of the institutionalised role of organisations is provided by "the rules of procedure"[161] that require an NGO representative to be invited to attend all the meetings of the Bureau of the Meeting of the Parties. This Bureau consists of government representatives elected by the Meeting of the Parties.[162]

The rules also state that all the meetings of the Convention bodies "shall be open to members of the public, unless the Meeting of the Parties, in exceptional circumstances, decides otherwise".[163] The representatives of "relevant non-governmental organisations, qualified or having an interest in the fields to which the Convention relates," are automatically "entitled to participate in the

---

159 See also: Cramer, 2009, p. 95.

160 For an NGO perspective, see Petkova, & Veit, 2000; Morgera, 2005, p. 140.

161 Rules of Procedure, Decision I/1 of the first Meeting of the Parties, Doc. ECE/MP.PP/2/Add.2, 2 April 2004, preamble.

162 Rule 22, § 2.

163 Rule 7, § 1.

proceedings of any meeting governed by these rules, unless one third of the Parties present at that meeting objects to the participation of representatives of that organisation".[164]

An NGO does not need a formal accreditation procedure but it just simply needs to express its wish to attend. Although the NGO representatives generally only have observer status without the right to vote, they do have the right to speak. In general, the President invites "speakers in the order in which they signify their desire to speak, but may at his or her discretion decide to call upon representatives of Parties before observers".[165]

The rules of procedure also stipulate that all official Convention body documents must be "placed on the ECE web site when sent to the Parties" and "provided to members of the public on request".[166]

1.3.2.2 Notion of "Public" and of "Public Concerned"

The Aarhus Convention uses two terms to define who has been granted rights: "the public" and "the public concerned".

Article 2(4) defines "public" without a reference to citizenship and applies the "any person principle". There is an extremely wide definition of members of the "public" as comprising all "natural or legal persons, and [...] their associations, organisations or groups". This definition covers a much broader spectrum than just the NGOs and even goes beyond Agenda 21's notion of "major groups". It encompasses the entirety of civil society and the private sector (Pallemaerts, & Moreau, 2004, p. 183). This term is not subject to any conditions. Thus, the question of whether a determined member of the public is affected or has an interest is not important where rights under the Convention are granted to the public.

Further, Article 3(9) stipulates that no person can be excluded from such definition on the grounds of nationality, domicile, citizenship, or place of registered seat.[167] Indeed, this non-discrimination clause is, according to the Implementation Guide, "the key provision of the Convention" (Stec, & Casey-Lefkowitz, 2000, p. 48). Indeed, Article 3(9) recognises that national borders do not constitute impermeable environmental barriers by insisting that the

---

164   Rule 6, § 2.

165   Rule 27, § 1.

166   Rule 11. Pallemaerts, & Moreau, 2004, p. 187.

167   For example in cases where the area potentially affected by an activity crosses an international border, members of the public in the neighbouring country might be members of the public concerned for the purposes of Art 6.

prescribed rights apply to all persons without discrimination as to citizenship, nationality or domicile.

This is a particularly interesting norm since it challenges the conventional idea that the "*demos*" is made up of citizens of a single state (Petkova, & Veit, 2000). Of course, the *demos* or public which has opportunities to participate, as provided by Aarhus, is still made up of the citizens of a single state. The novelty lies in the fact that the people affected had the right to participate in environmental decisions that have crossed boundaries.[168] Consequently, these provisions open the door to an environmental citizenship, at least concerning the access to information. An example of this approach is the rights under Article 4, which relate to requests for information; this article applies to non-citizens and non-residents as well as citizens and residents.

Additionally, the Convention's definition of public differs from that of other Conventions, also concerning the language referring to associations, organisations or groups of natural or legal persons.

The words have been interpreted by the Implementation Guide "to provide that associations, organisations or groups without legal personality may also be considered to be members of the public under the Convention" (Stec, & Casey-Lefkowitz, 2000, p. 39).

The term "public concerned" provided for in Article 2(5), also underlined by the Implementation Guide, makes allusions to "a subset of the public at large that has some kind of a special relationship to a particular environmental decision-making procedure. To be a member of the public concerned in a particular case, the member of the public must be likely to be affected by the environmental decision making, or the member of the public must have an interest in the environmental decision-making".

The above mentioned expression can also be found in Article 6 on public participation in decisions on specific activities, as well as in the linked access-to-justice provisions of Article 9(2).

Although less comprehensive than "public", the scope of the term "public concerned" is still quite extensive. It should be noted that it seems "to go well beyond the kind of language that is usually found in legal tests of 'sufficient interest', including not only the members of the public who are likely to be affected, but also the members of the public who have an interest in the environmental decision-making" (Stec, & Casey-Lefkowitz, 2000, p. 40-41).

---

168   Some scholars are certain that "the mobilization of local actors across borders" is an appropriate response to the "processes of economic globalization". See: Petkova, & Veit, 2000.

It also applies to a group of the public, which has an indeterminate interest in the decision-making procedure. Indeed, Article 2(5) does not demand that a person must demonstrate a legal interest to fall within the definition of the "public concerned". Thus, the Convention seems to allow an equal status, at least concerning procedural rights and potential remedies, regardless of whether the interest is a legal or factual one. [169]

The Aarhus Convention automatically qualifies, for the purpose of participation in environmental decision-making, NGOs as falling within the term "public concerned". Thus, they must neither prove a specific interest, nor demonstrate that they are affected or likely to be affected by the environmental decision-making, as long as they meet the requirements under national law.

States can lay down conditions for NGOs under national law, but these conditions should be in accordance with the Convention, and must therefore meet requirements such as non-discrimination and prevention of technical and financial obstacles to registration. Furthermore, once an NGO has fulfilled the requirements, it constitutes a member of the "public concerned" with regard to any purpose within the Convention, and also may possess a sufficient interest according to Article 9(2). [170]

2 Environmental Rights and Ecological Duties in International Environmental Law and the Aarhus Convention

This part will study whether in international environmental law and in the Aarhus Convention an environmental and/or ecological approach can be identified, as studied in the first Chapter; in other words, if it is already possible to affirm that

---

169   Generally, in national law persons with a mere factual interest do not normally enjoy the full panoply of rights in proceedings and judicial remedies accorded to those with a legal interest under these systems. Such as those of Austria, Germany Italy and Poland. Stec, & Casey-Lefkowitz, 2000, p. 39.

170   But it has been remarked for NGOs that do not meet such requirements *ab initio*, and for individuals, the Convention is not entirely clear whether the mere participation in a public participation procedure under Article 6, paragraph 7, would qualify a person as a member of the "public concerned". Because Article 9, paragraph 2, is the mechanism for enforcing rights under Article 6, however, it is arguable that any person who participates as a member of the public in a hearing or other public participation procedure under Article 6, paragraph 7, should have an opportunity to make use of the access-to-justice provisions in Article 9, paragraph 2. In this case, he or she would fall under the definition of public concerned. Stec, & Casey-Lefkowitz, 2000, p. 40-41.

there is a kind of recognition of environmental and ecological citizenship and its corresponding rights and duties.

What one can already reveal is that international law generally does not provide for environmental rights and ecological duties, first because there is no internationally binding environmental rights treaty and second, even if certain treaties refer to them, those are classified as "soft law" and consequently are not enforceable. Despite the aforementioned situation, the Aarhus Convention has made tangible progress, in particular in affirming an environmental citizenship granting environmental procedurals rights.

The three pillars and the two additional sections which form the structure of the Aarhus Convention are essential to the achievement of the right to a healthy environment, as well as the possibility of individuals fulfilling their responsibilities towards others, especially including future generations (Stec, 2003).

2.1 Environmental Rights: the Link Between the Environment and Human Rights in International Law

At the international level, the link between human rights and environmental degradation began to be realised as early as 1968 when the General Assembly of the UN[171] recognised that "impairment of the environment could have a direct effect on a person's enjoyment of basic human rights". Later other international human rights bodies[172] started to underline the connection (Soveroski, 2007, p. 261).

One of the most eloquent statements of support for a link between environmental protection and human rights has been made by former Vice-President Christopher Weeramantry's in his separate opinion in the Gabcíkovo-Nagymaros case before the International Court of Justice: "The protection of the environment is likewise a vital part of contemporary human rights doctrine, for it is a sine qua non for numerous human rights such as the right to health and the right to life itself. It is scarcely necessary to elaborate on this, as damage to the environment can impair and undermine all the human rights spoken of in the

---

171   The UNGA in G.A. Res. 2398 (XXII).

172   The Special Reporters to the former United Nations Human Rights Commission on housing, health, indigenous peoples' rights, and migrants have all stressed the connectedness of environmental protection and human rights to their areas of study and review. See: Sensi, 2004, p. 6.

Universal Declaration and other human rights instruments. While, therefore all peoples have the right to initiate development projects and enjoy their benefits, there is likewise a duty to ensure that those projects do not significantly damage the environment".[173]

Besides this recognition, the protection of the environment is justified on the grounds of its importance to the enjoyment of basic human rights and human survival, and it becomes clear in his statement that environmental rights have expanded so much they damage the environment (Pedersen, 2010).

An example of this link has been also reaffirmed in a recent resolution adopted by the UN Commission on Human Rights (CHR) entitled "Human rights and the environment as part of sustainable development",[174] when its preamble merely recalls "that environmental damage can have potentially negative effects on the enjoyment of some rights" (Bjerler, 2009).

Little by little, a myriad of declarations, international conventions and agreements that address human rights and environmental protection have reflected the international community's recognition that international actors, particularly States, have obligations in these areas, and individuals, as well as groups of peoples, have a number of environmental rights.

Although all declarations, resolutions and reports remain very important on a political level, they are not binding obligations. Consequently, it is necessary to look at soft law to find a legal foundation of environmental rights, even if based on a non-binding foundation (Bjerler, 2009).

2.1.1 Substantive Environmental Right

As early as 1974, the Nobel Prize winner Cassin (Cassin, 1974) was the first who spoke of the substantive "right to the environment" and the possibilities for its development in the future. Such a view was not isolated; Gormley, in his book "Human Rights and Environment: the Need for International Cooperation" (Gormley, 1976), in 1976 described the connection between human rights

---

173 ICJ Judgment 25 September 1997. Already in 1995 the Court's statements reflect a growing recognition amongst some of the judiciary at the ICJ to bring new environmental principles into the reasoning of their judgements. The ICJ heard a case brought by New Zealand relating to the intention of France to carry out nuclear tests in the Pacific Ocean. New Zealand requested inter alia that the court order France to carry out an EIA in accordance with international law. It further argued that such tests would be illegal unless the ICJ declined jurisdiction; however, three dissenting opinions did address the maters concerned. Judge Weeramantry found that there was a prima facie obligation on France to carry out an EIA and he also referred to international support for the precautionary principle and intergenerational equity: Pedersen, 2010.

174 CHR Res. 2003/71 adopted on 25 April 2003.

and the environment and he affirmed that a substantive environmental right already existed: "the mere physical existence of man – as guaranteed in national constitutions, the Universal Declaration of Human Rights, the United Nations Charter, and the European Convention of Human Rights and Fundamental Freedoms – demonstrates all too clearly the need for greater recognition of the new human right to a minimum decent environment" (Horn, 2004, p. 233).

Although the establishment of an "environmental right" is recognised by almost all scholars as indispensable for the enjoyment of other human rights and freedoms, the recognition of it suffers from several problems.

There is a problem of definition: the complexity of environmental processes and the great variety in environmental circumstances make it very difficult to answer the question on the content and scope of a right to "the environment". Adding a qualifying adjective such as "decent", "healthy" or "viable" does not solve this deficiency either.

Thus, if an environmental right is to be more than a rhetorical slogan it implies the promotion of a certain level of environmental quality. However, it has been underlined by Kiss that the vagueness "is not exceptional in the field of human rights, where concepts such as 'order public', 'national security', 'morality' are to be given an exact interpretation" (Kiss, 1992, p. 201).

The right to environment similarly can be interpreted, not as the right to an ideal environment, difficult if not impossible to define in abstract, but as "the right to have the present environment conserved – protected from any significant deterioration – and improved in some cases" (Kiss, 1992).

Accordingly, a substantive right to environment would essentially be of a programmatic nature, and hence require progressive realisation on behalf of the State to fulfil such a right, subject to the availability of sufficient resources. Thus, it has been argued by some commentators that the introduction of this new human right "may add little to the progress achieved already and could, in fact, hinder the development of international environmental law" (Boyle, 1996, p. 43; Pevato, 1989, p. 309).

Moreover, another problem is that in case of an introduction of such a substantive right, it is also necessary to provide satisfactory legal mechanisms of enforcement (Bjerler, 2009); besides, an essential aspect of protecting human rights is the creation of procedures that permit ensuring respect for these rights.

The introduction of this right has also been criticised since it would be too anthropocentric insofar as it is only for the benefit of humans (Giagnocavo, & Goldstein, 1990, p. 356-357). Despite that vision, other commentators have argued that this right could be an example of "weak anthropocentrism" (Redgwell, 2007, p. 159) because, even if this right's focus remains human benefit, the concept is drawn so broadly as to be indistinguishably eco-centric (Birnie, Boyle, & Redgwell, 2009, p. 280).

In fact, the recognition of this right would be less anthropocentric than the present law: "it would benefit society as whole, not just individual victims. It could enable litigants and NGOs to challenge environmentally destructive or unsustainable development." Moreover, although the interests of humans are at present protected primarily, the addition of the right to a clean environment could be seen as complementary to this wider protection of the biosphere, "reflecting the impossibility of separating the interests of mankind from the environment as a whole, or from the needs of future generations" (Birnie, Boyle, & Redgwell, 2009, p. 280).

It has been said that in general most of the international literature calls for an express recognition of such a right in human rights documents and in international environmental law. A recent development towards this path was made in February 2007: an appeal was adopted at a conference bringing together NGOs, senior United Nations officials and the French head of state. This appeal stated "to promote environmental ethics, we are calling for the adoption of a Universal Declaration of Environmental Rights and Duties. This common charter will ensure that present and future generations have a new human right to a sound and well-preserved environment".[175]

In conclusion it can be said that currently there is no international treaty or agreement that provides a globally accepted substantive human right to protect the environment.[176] Against this background some scholars have affirmed that in the field of international environmental law, non-binding approaches have assumed a substantially greater importance and thus the numerous soft law provisions addressing the human right to environment have "more than mere rhetorical force. Rather, they are the likely precursors to binding international legal obligations in this area" (Collin, 2007a, p. 125).

For this reason, a glance at soft law shall be made.

2.1.1.1 Substantive Right in International Soft Law

A trend in international law to incorporate such a "decent environmental" right within the environmental law regime can be observed in several documents.

Although not explicitly mentioned, the inspirational language of the Stockholm Declaration has often been heralded as the first international proclamation of a human right to a clean environment (Sohn, 1973, p. 455).

---

175 Paris Appeal, 2007, see: P. Taylor, 2009, p. 96.
176 Example Article 21 of Rio Declaration is widely regarded as reflection customary international law; Sands, 2003, p. 146.

This declaration stated that: "Man has the fundamental rights to freedom, equality and adequate conditions of life in an environment of a quality that permits a life of dignity and well-being".[177]

Later, the Brundtland Report, entitled Our Common future, published in 1987, followed this view: the World Commission on Environment and Development adopted a catalogue of legal principles for environmental protection and sustainable development, the first of which reads: "All human beings have the fundamental right to an environment adequate for their health and well-being" (Déjeant-Pons, 2002, p. 23).

Despite this initial emphasis on a human rights perspective, twenty years after the Stockholm Conference, at the Rio Conference, this prospective was not maintained (Shelton, 1992, p. 82). Avoiding the terminology of rights altogether, the Rio Declaration merely asserted that: "Human beings are at the centre of concerns for sustainable development. They are entitled to a healthy and productive life in harmony with nature".[178]

According to some authors, the Rio Declaration's failure to award a larger enunciation to human rights is symptomatic of the continuing indecision and debate relating to the appropriate place of human rights law in the development of international environmental law (Shelton, 1992, p. 82; Boyle, 1996, p. 43), and it was also argued that the traditional international legal interpretation of human rights is inadequate for these rights to protect the environment.

This pushed the UN Sub-Commission on the Prevention of Discrimination and Protection of Minorities to embark on a study on human rights and the environment. The final Report of 1994 also called Ksentini Report, proposes a notion of human rights and their interaction with the environment.[179] Moreover, the Sub-Commission goes on to suggest a declaration of "Principles on Human Rights and the Environment".

These principles proclaim that: "all persons have the right to a secure, healthy and ecologically sound environment", and that "all persons have the right to an environment adequate to meet equitably the needs of present generations and

---

177 Declaration on the Human Environment, Principle 1, Report of the United Nations Conference on the Human Environment, New York 1973.

178 Declaration on Environment and Development, Principe 1, Report of the UN Conference on Environment and Development (UNCED). The UNCED was held in Rio de Janeiro (Brazil) from 3 to 14 June 1992 and was attended by 178 States, more than 50 intergovernmental organisations and several hundred non-governmental organisations (NGOs). The European Union also attended the Conference. In addition to the signing by more than 150 States of the United Nations Framework Convention on Climate Change and the Convention on Biological Diversity, the Conference adopted three non-binding instruments: the Rio Declaration, the UNCED Forest Principles and Agenda 21.

179 Ksentini Report: Final Report, UN Doc. E/CN.4/Sub.2/1994/9.

that does not impair the rights of future generations to meet equitably their needs"(Boyle, 1996, p. 43; Hayward, 2000, p. 558).

In addition, the revolutionary approach of this Report, which is collocated between the anthropocentric and eco-centric prospectives, may be noted. Sure enough, one of the aims of this document was to ensure that human rights law also covers the protection of the environment (Anderson, 1996, p. 22). Indeed, in terms of human rights, these rights include protection and preservation of flora and fauna, essential processes and areas necessary to maintain biological diversity and ecosystems, as well as conservation and sustainable use of nature and natural resources. Drawn in such a broad way, their substance is far from being exclusively anthropocentric.

In 2002, the World Summit on Sustainable Development[180] which was supposed to be the natural follow-up to the Stockholm and Rio Conferences, did not produce a set of principles similar to the previous conferences (Kiss, & Shelton, 2004; 2007; Marks, 2004, p. 137) and the relevance of human rights was only mentioned in the Plan of Implementation; thus, "the development of environmental rights was not taken further" (Turner, 2009, p. 9; Gormley, 1976).

Moreover, it should be noted that some scholars claim that a human right to a healthy or clean environment has progressively developed and is still developing; and it can also be said that this reflects in its implementation *via* national constitutions, regional treaties[181] or in case law.[182]

---

180   Johannesburg Declaration on Sustainable Development adopted by the World Summit on Sustainable Development, 2-4 September 2002, U.N. Doc. A/CONF.199/20, §13.

181   See African Charter on Human and Peoples' Rights, adopted by the Organisation of African Unity, 27 June 1981, entered into force 21 October 1986, Article 24; International Covenant on Economic, Social and Cultural Rights (ICESCR), Article 12; Convention on the Rights of the Child, adopted by G.A. Resolution 44/25, 20 November 1989, entered into force 2 September on 1990, Article 24 (2) (c); European Social Charter, adopted by the Council of Europe, 18 October 1961, entered into force 26 February 1965, Article 11; Additional Protocol to the American Convention on Human Rights in the Area of Economic, Social and Cultural Rights, "Protocol of San Salvador", adopted by the Organisation of American States, 17 November 1988, entered into force 16 November 1999, Article 11.

182   South African courts also have deemed the right to environment to be justifiable. In Argentina, the right is deemed a subjective right entitling any person to initiate an action for environmental protection. Colombia also recognises the enforceability of the right to environment. In Costa Rica, a court stated that the right to health and to the environment are necessary to ensure that the right to life is fully enjoyed. See Shelton, 2007, p. 26.
In India, for example, a series of judgments between 1996 and 2000 answered to health preoccupations produced by industrial pollution in New Delhi. Razaque, 2002. As early as 1991, the Supreme Court interpreted the right to life guaranteed by Art. 21 of the Constitution to include the right to a whole some environment. In a subsequent case, the Court observed that Article 21's guarantee of right to life includes the right to enjoy pollution-free water and air. Jona Razzaque has underlined also the jurisprudence

In particular, many constitutions have changed throughout the last twenty years and they have been amended to specifically accommodate environmental rights. Now, several hold provisions on substantive as well as procedural environmental rights and this adds further impetus to the use of rights to provide for environmental protection.[183] Moreover, the rights in the national constitutions have the potential to influence debates on the status of a substantive environmental norm under international law, since they can constitute strong indications of national *opinio juris.*

Finally, it may be argued that the "abundance" of state action in the form of national and international environmental laws, including constitutional norms granting the right to environment, provide "strong evidence of the emergence of the right to environment as a principle of customary international law" (Collins, 2007a, p. 129).

---

from Bangladesh and Pakistan. Those above mentioned countries don't provide express constitutional rights to an adequate environment but instead their judiciaries have used various existing constitutional rights to protect the environment. In particular the right to life has been extended to include the right to a healthy environment (Case of Kendra in 1985: the Indian Supreme Court had alluded "to the right of people to live in a healthy environment with minimal disturbance of the ecological balance". In 1990, "the link between environmental quality and the right to life was made explicit by a constitution bench of the Supreme Court in Charan lal v. Union of India. In 1991 the Court interpreted the right to life provided by the Article 21 of the Constitution to "include the right to a clean environment".

In the 1994 case of Dr. M. Farooque v. Bangladesh, the Supreme Court declared that the constitutional right to life extends to include the right to have a healthy environment. Also in Pakistan the right to a clean environment has been established by the Pakistani judiciary). Moreover, the single most important case to date is from the Philippines in 1990, Oposa Minors, regarding the concept of intergenerational responsibility (Hayward, 2005, p. 208-209).

183 Thus, today many constitutions throughout the world guarantee a substantive right. Research undertaken by Earthjustice, published in 2005, has found that "of the approximately 193 nations of the world there are now 117 whose national constitutions mention the protection of the environmental natural resources. 109 of them recognise the right to a clean and healthy environment and/or the State's obligation to prevent environmental harm. Of these, 56 constitutions explicitly recognise the right to a clean and healthy environment, and 97 constitutions make it the duty of the national government to prevent harm to the environment. Fifty-six constitutions recognise a responsibility of citizens or residents to protect the environment, while 14 prohibit the use of property in a manner that harms the environment, or encourage land use planning to prevent such harm. Twenty constitutions explicitly make those who harm the environment liable for compensation and /or remediation of the harm, or establish a right to compensation for those suffering environment injury. Sixteen constitutions provide an explicit right to information concerning the heath of the environment or activities that may affect the environment". See Mollo, 2005, pp. 37-38; Turner, 2009, p. 27-28; May, 2005-2006, p. 113.

2.1.1.2 Substantive Right in the Aarhus Convention

Principally, the Convention's prime concern is not the substantive right to a healthy environment, but the procedural rights of access to information, access to decision-making and access to justice. Nevertheless, to better understand the Convention, it is useful to start by examining said right.

To analyse the substantive right to a healthy environment as contained in the Convention it is necessary to begin by briefly analysing the history of the Aarhus Convention.

In this respect, the Ministerial Conference on Sustainable Development in the ECE Region, a regional preparatory meeting for UNCED held in Bergen, Norway, in May 1990, played an important role. There, member States of the UNECE agreed to contribute to the drafting of a document on environmental rights and obligations for possible adoption at the 1992 Conference on Environment and Development.[184]

Nevertheless, the first version of the Charter on Environmental Rights and Obligations, which clearly acknowledged the right of the individual to a healthy environment as well as procedural rights, was never officially approved by the UNECE, also due to the fact that the provision recognising the existence of a substantive right to the environment was particularly debated.[185]

Thus, the UNECE preferred to put exclusive attention on the implementation of the procedural rights as established in Principle 10 of the Rio Declaration. However, following a suggestion proposed at the first meeting of the working group by the Belgian delegation, it was agreed to incorporate at least in the Preamble a reference to the substantive right to a healthy environment (Pallemaerts, 2002, p. 17).

The Preamble, hence, recognises that: "*adequate protection of the environment is essential to human well-being and the enjoyment of basic human rights, including the right to life itself*" and "*also that every person has the right to live in an environment adequate to his or her health and well-being*".

The explicit acknowledgment of the right to a healthy environment adds weight to its operative provisions for the implementation of procedural rights.[186]

---

184   Bergen Ministerial Declaration on Sustainable Development in the ECE Region, 16 May 1990, paragraph 16.

185   See Report of the Ad hoc Meeting on environmental Rights and Obligations, UN Dc. ENVWA/AC.7/2 10 July 1991, par. 10.

186   Collins affirms: "As the above survey of existing provisions makes clear, the emerging right to environment has been modified by a variety of adjectives: clean, healthy, ecologically balanced, sound, healthful, adequate, viable, decent, sustainable, etc. Given that the environmental human rights agenda has developed as a response to environmental harm, it is helpful in analyzing this question to recall that such harm generally

As it has been stressed, the meaning of the above-mentioned sentences shows "that they are not ends in themselves, but are meaningful precisely as means towards the end of protecting the individual's substantive right to live in a healthy environment" (Pallemaerts, 2002, p. 18).

The Preamble does not have immediate legal consequences or entail any specific obligations on parties beyond those laid down in the other provisions of the Convention. Moreover, some States, to ensure that the promise written in the preamble would be not binding, explicitly excluded such a meaning. For example, the United Kingdom made a declaration while signing the Convention, underlining that it would "interpret the above provision as "an aspiration" and not a legal right" (Turner, 2009, p. 9).

Against this interpretation some authors have stated: the Convention is recognising a human right to "live in an environment adequate to (one's) health and well-being" and some authors have said that this human right "constitutes the core of the Aarhus conception of sustainability".[187] Also, the Implementation Guide confirms this interpretation: "this proclamation, although merely contained in the Preamble, is the first explicit recognition of the right to

---

falls into one of two interconnected categories: i) contamination (including for example, radiation, the contamination of drinking water sources with hazardous chemicals, and air pollution) and ii) the destruction of natural habitats (encompassing the loss of biodiversity, wilderness, aesthetic values, and eco-cultural spaces). As a result, this author supports the adoption of the modifier "healthy and ecologically balanced" in discussions of the right to environment. The term "healthy" responds most directly to environmental contamination causing direct human health effects, while "ecologically balanced" responds to the second category of environmental harm. As noted at the outset, there is a live debate regarding the utility and desirability of an international human right to environment, because of the concern that "to speak of a human right to a healthy environment detracts from [an] ecocentric approach to environmental protection and, instead, endorses the rather narrow [...] anthropocentric approach". The phrase "healthy and ecologically balanced" responds to this concern as the health component addresses human-centered needs directly, while the phrase "ecologically balanced" is an eco-centric concept consistent with the notion of "ecosystem integrity", which has been suggested as the basis for a new human relationship with nature. The formulation "healthy environment" has been widely used, and should in any event be understood as encompassing both human-centred and eco-centric aspects– as in an environment that is both "healthy" for humans and healthy in its own right (e.g. a healthy lake, a healthy forest, a healthy ecosystem). Collins, 2007a, p. 137.

187 Moreover it has been underlined that this kind of sustainability is negative (About ‚negative' conceptions of sustainability see Meadowcroft, 1997, p. 167). "This means that the "requirement is an environment adequate, it is only if an environment is not adequate that it is unacceptable. There may be many different states of the environment that meet the minimum standard of being adequate to health and well-being". "The range of adequate states of the environment will depend on how we define ‚health and well-being and the larger present and future populations of the earth, the narrower the range of adequate ‚states of the environment'". Bell, 2004, p. 96.

a healthy environment in an international instrument" (Stec, & Casey-Lefkowitz, 2000, p. 16).

Furthermore, Article 1 instructs Parties how to move towards the establishment of the guarantee of the basic right of present and future generations to live in an environment adequate to health and well-being. Thus, Article 1 affirms: "In order to contribute to the protection of the right of every person of present and future generations to live in an environment adequate to his or her health and well-being, each Party shall guarantee the rights of access to information, public participation in decision-making, and access to justice in environmental matters in accordance with the provisions of this Convention".

It may be noted that the literal meaning of Article 1 is that the parties acknowledge that guaranteeing the rights laid down in the convention will not in itself be sufficient to ensure the protection of the substantive right, but will only contribute to "the achievement of that ultimate objective" (Pallemaerts, 2002, p. 18). This norm establishes "the linkage between practical, easily understandable rights, such as those relating to information and decision-making, and the harder-to-grasp complex of rights included in the right to a healthy environment" (Stec, & Casey-Lefkowitz, 2000, p. 29).

To sum up, the protection of the right to a healthy environment is presented in inspirational terms as an objective to which the Aarhus Convention is intended to contribute, not as a "substantive obligation distinct from the specific obligations with respect to access to information, participation and access to justice which it imposes on its contracting parties" (Pallemaerts, 2004, p. 18). On the basis of the Convention, which guarantees in theory the procedural rights of present generations, "not only is the substantive right of future generations to live in a healthy environment protected but also the rights of participation in decision-making which are a precondition for the enjoyment of the former" (Nadal, 2008, p. 28).

So the substantive right finds itself in procedural dimensions. In this respect there is in fact a close affinity between the Convention and international human rights law. This affinity also appears from the Convention's provisions on compliance review, which will be explored later, that for the first time in international environmental law, "opened up the possibility of establishing a review mechanism accessible not only to states, but also to individuals, through some form of individual recourse procedure".[188]

---

[188] The seventh and eighth preambular paragraphs also recognize the contained procedural rights and duties of the Convention as a precondition for the enjoyment of the right to a healthy environment. Pallemaerts, 2004, p. 14.

2.1.2 Procedural Environmental Rights

It has been seen in the first Chapter that no matter how strong a substantive right to a clean environment might be on paper, it would be meaningless without the procedural rights necessary to pursue respect, protection and promotion of that right.

There are several advantages to introducing them. First, development and use of environmental procedural rights will not only provide opportunities to protect environmental rights but can also further the development of a substantive right to a clean environment (Soveroski, 2007, 261).

Second, they can overcome the difficulties with the definition of the right to a healthy environment and its enforcement. These rights establish where and how substantive rights can be enforced, and against whom. These rights would be implemented by individuals to enable them to enforce rights to acquire information about acts of environmental degradation, and the right to be able to participate in the making of decisions affecting the environment and to have access to administrative and judicial remedies.

Finally, procedural rights can also be employed to avoid the problem of anthropocentricity by including protection of nature's interests, such as by allowing NGOs to take actions on behalf of the environment.

2.1.2.1 Procedural Environmental Rights in International Environmental Law

Thus, in order to ensure the most effective and immediate upholding of a right to a clean environment it may be more expeditious to focus on further development and expansion of procedural, rather than substantive right.

In comparison to substantive rights, the mechanisms which procedural rights depend on are more politically acceptable and have acquired important international support, not only in soft law but also in binding treaties.[189] These rights simply give practical form, in the specific context of environmental policy, to the general principles of democracy and the rule of law already enshrined in other international instruments on the protection of human rights.

---

189   For Articles on public participation in international environmental contexts, see Becker, 1993, p. 235; Breitmeier, & Rittberger, 2000, p. 130; Cameron, 1996, p. 29; French, 1996, p. 251; Hayton, 1993, p. 275; Hostetler, 1995, p. 279; Peel, 2001, p. 47; Powell, 1995, p. 109; Raustiala, p. 537; Sanchez, 1993, p. 283; Sands, 1991; Spaulding, 1995 p. 1127; Waak, 1995, p. 345.

That procedural environmental approach is supported and has been promoted by international instruments such as the 1982 World Charter for Nature and the Rio Declaration amongst others.[190]

According to Paragraph 23 of the World Charter for Nature: "All persons, in accordance with their national legislation, shall have the opportunity to participate, individually or with others, in the formation of decisions of direct concern to their environment, and shall have access to means of redress when their environment has suffered damage or degradation".[191]

Nevertheless, the most important statement on environmental procedural rights is contained in the Rio Declaration: "Environmental issues are best handled with the participation of all concerned citizens at the relevant level".

And in particular Principle 10: "*Each individual shall have appropriate access to information concerning the environment that is held by public authorities, including information on hazardous materials and activities in their communities, and the opportunity to participate in decision-making processes. States shall facilitate and encourage public awareness and participation by making information widely available. Effective access to judicial and administrative proceedings, including redress and remedy, shall be provided*".

The participatory norms embodied in Principle 10 have, more than many aspects of International Environmental Law, challenged traditional ideas and limits of both international law and municipal law (Anton, 2008, p. 8). In other words, this norm appears to open up areas of a state's *domain réservé* (Anton, 1993, p. 553). The inward bearing of these international participatory norms has the potential to invade fundamental aspects of the state including state secrecy, legal procedure, and public administration.

Indeed, the norms elaborated in Principle 10 have been seen by some commentators as an attempt to break the external "sovereignty barrier"[192]

---

190   Agenda 21 also indicates that "individuals, groups and organisations should have access to information relevant to environmental rights could be based upon current political rights together with the development of the right of the individual to take action to protect the environment".

191   See World Charter for Nature (UN Doc. A/RES/37/7, 28 October 1982).

192   The term „sovereignty barrier" is taken from an anon Article appearing in an NGO newspaper for the 1992 United Nations Conference on Environment and Development (UNCED). Sovereignty Barrier to environmental Law, Crosscurrents, August 14-15, 1991, p. 2, col. 1. Alan James has characterized the external dimension of sovereignty thus: "[F]or the Solomon Islands and Tuvalu, as for all other internationally active states, the sovereignty on which their international activity is based amounts to constitutional separateness. A sovereign state may have all sorts of links to other states and international bodies, but the one sort of link which, by definition, it cannot have is a constitutional one. For sovereignty […] consists of being constitutionally apart, of not being contained, however, loosely, within a wider constitutional scheme". James, 1986. This external

by granting participatory rights on individuals and other non-state actors in fundamental aspects of the international system such as international law-making and monitoring compliance with and enforcing breaches of international law (Anton, 2008, p. 8; Donald, 2008).

In other words, this norm has tried to stir up the basis of environmental governance and introduce a participatory element of the environmental democracy model.

2.1.2.2 Procedural Environmental Rights in the Aarhus Convention (see Section II of Chapter II)

The most important binding instrument in which procedural rights are recognised is the Aarhus Convention. As has already been underlined, this Convention emphasises connections between "environmental goals, participative democracy and individual rights, all of which are components of the current interest in environmental democracy" (Steele, 2001, p. 415).

The second paragraph in the Preamble of the Aarhus Convention recalls Principle 10 of the Rio Declaration, which drew attention to the creation of new procedural rights. These were to be granted to individuals through international law and exercised on a national and possibly international level (Stec, & Casey-Lefkowitz, 2000, p. 13; Sands, 1995, p. 99).

The Aarhus Convention implemented Principle 10, and as the Convention's full name suggests, in the form of three pillars: access to information, public participation and access to justice.

After looking briefly at the general content of those pillars in this section, each will be discussed in more detail in the next section.

Access to information constitutes the first of the three pillars. This is due to the fact that effective public participation in decision-making relies on full, precise and updated information. It is also able to stand alone, since the public may ask for access to information for any reason, not just for the purpose of participating. The access to information pillar is divided into two parts.

The first type of access to information is *passive* and is covered by Article 4. It relates to the right of the public to *receive* information from public authorities, as well as the obligation of the public authorities to give information after a submission. Article 5 covers the second type, called *active* access to information (Stec, & Casey-Lefkowitz, 2000, p. 4). It involves the right of the public to

---

constitutional independence, or separateness, based on the characteristics of statehood is what provides states with legal personality on the international plane and the exclusive ability to create international law, including law that allows or prohibits non-state actors from participating in the international legal system. Arend, 1999, p. 86-87.

obtain information and the obligation of authorities to collect and disseminate information of public interest without the necessity of a precise request. The recognition of this right reflects the deliberative and participatory theories in which the *informed* citizen is seen as a step closer to awareness and participation than the uninformed.

Public participation constitutes the second pillar of the Aarhus Convention. It is split into three parts. The first part is related to the participation of the public, as provided for in Article 6; that is, decision-making concerning a definite activity by which the public can be affected or which, for any other reason, the public is interested in. The second part concerns the participation of the public in the development of plans, programs and policies relating to the environment, and is provided for in Article 7.

Finally, Article 8 provides for participation of the public in the preparation of laws, rules and legally binding norms (Stec, & Casey-Lefkowitz, 2000, p. 4). This pillar is in the centre of the Convention. It encompasses the deliberative as well as participatory democracy approach, since it utilises especially the participatory mechanisms, but also introduces one deliberative mechanism, which will be studied below.

Article 9(1) and (2) covers the third pillar of the Convention, access to justice. It deals with access to justice in two situations: first, it protects the other two pillars, access to review procedures in relation to information and access to review procedures to challenge decisions, acts, or omissions subject to the public participation provisions of Article 6. Secondly, because it helps fulfil the duty of protecting the environment for future generations. Article 9(3) has been considered the fourth pillar of the Aarhus Convention, because it provides access to administrative or judicial procedures to challenge acts and omissions by private persons and public authorities which breach environmental law (Marshall, 2006, p. 126).

Finally, there is the fifth pillar which grants to the individual access to a Review Mechanism, by participation in monitoring of state compliance with Aarhus legal obligations.

2.2 Ecological Duties

As seen in the first Chapter, the concept of ecological duties has been introduced over the last 60 years by debates within environmental philosophy and prevailing environmental ethics. In fact, while environmental rights appear as legal entitlements, ecological duties are, at best, referred to as moral obligations. Consequently, these duties are not binding at the level of international law.

Nevertheless, ecological approach values and ecological duties have only recently emerged in legal traditions. The next part briefly considers some of the precedents of this ecological way in international law, in particular in the Aarhus Convention.

2.2.1 Ecological Duties in International Environmental Law

None of the human rights norms provides duties vis-à-vis future generations and vis-à-vis the environment, just a few timid attempts have been made at the international level, and generally this has occurred using "soft law" mechanisms.

Since the Stockholm Conference, international environmental law has stressed that humans have a "solemn responsibility to protect and improve the environment for present and future generations". This was the first time that an international declaration held the connection between the right to and the responsibility for the environment and this novelty most probably reflected the political climate of the time. Indeed, the experience of the environmental crisis was "fresh in the mind of the public and conference delegates, leading to a covenant-type declaration of rights and responsibilities" (Bosselmann, 2008,p.121).

A second essay to introduce the duties approach was the World Charter for Nature.[193] The Charter, which was the first international document to introduce an eco-centric view, states that mankind is responsible for all species, and promulgated provisions for fulfilling this responsibility.

It required that, according to Article 1 "Nature shall be respected and its essential processes not impaired", and Article 2 declared "the genetic viability on the earth shall not be compromised; the population level of all life forms, wild and domesticated must be at least sufficient for their survival, and to this end necessary habitats shall be safeguarded" (Birnie, Boyle, & Redgwell, 2009, p. 603). Thus, essential ecological processes and life-support systems must be maintained in the interest of subsistence as well as of the diversity of living organisms (P. Taylor, 2009, p. 104).

Paragraph 24 introduces, moreover, a kind of ecological citizenship when it states: "Each person has a duty to act in accordance with the provisions of the present Charter, acting individually, in association with others or through participation in the political process, each person shall strive to ensure that the objectives and requirements of the present Charter are met".

Unfortunately, the Word Charter is not legally binding and offers nothing more than general principles which are expressed in mandatory terms: for instance, the verb "shall" is used throughout rather than "should". A French commentator criticising this approach said that the Charter has "[one] apparence pseudo

---

193   UN GA RES 37/7: World Charter for Nature (1982).

juridique" adding that "il est à craindre que pour avoir voulu proposer du droit doux le législateur ne propose plus ici de droit de tout [...] pourquoi alors ce masque? Si cette pseudo-règle peut on espère servir la cause de la nature, elle ne peut que contribuer à discréditer celle du droit".[194]

Also, the Ksentini Report tried to undertake the duties path putting the accent not just on environmental rights, but also the reciprocal relationship between rights and duties with respect to the environment. For example Article 21 states, "All persons individually and in association with others, have a duty to protect and preserve the environment".

In 2000, the International Covenant on Environment and Development (IUCN) made another attempt.[195] This document, which is only a draft and has no legal status, adopts a holistic approach to the environment recognising the interdependency and interconnectedness of ecosystems.[196]

The Covenants had two main objectives: first, to establish a framework for the development of fundamental environmental principles;[197] second, to introduce less of an anthropocentric dimension by considering the inclusion not "only of legal obligations for natural systems but also the possibility of granting legal rights to natural systems" (Horn, 2004, p. 233).

For the above mentioned reasons, Article 2 affirms: "Nature as a whole warrants respect; every form of life is unique and is to be safeguarded independent of its value to humanity" (P. Taylor, 2009, p. 104) Then, this Covenant affirms, there is "the duty of all to respect and care for the environment" and Article 12 specifies who has this duty, stating in paragraph 2: "Parties shall ensure that all natural and juridical persons have a duty to protect and preserve the environment".[198] In other words, this obligation has to be fulfilled by international organisations, states, business communities, associations and individuals. They must abstain from harm to the environment, "respect" and take affirmative action to "preserve", to protect, and, where necessary, rehabilitate it (Horn, 2004, p. 233).

---

194 Remond- Gouilloud, 1982, p. 120. Contrary to another point of view, some commentators have underlined that the Charter indicates the new direction of international environmental law. See Kiss, & Shelton, 1991, p. 46-48; P. Taylor, 2009, p. 104.

195 Draft International Covenant on Environment and Development, 1995 and update in 2000 and 2004, Cambridge

196 Draft International Covenant on Environment and Development 2000, (IUCN Commission on environmental Law and International Council of environmental Law); Burhenne, & Tarasofsky, 1998, p. 77.

197 Congress on Public International Law, (1995), p. 163.

198 For some commentators the term "duty" here expresses a legal obligation and not only a moral one. See Horn, 2004, p. 233.

2.2.1.1 The Earth Charter

One of the most developed ecological approaches can be found in the provisions of the Earth Charter. This Charter is unique in its efforts to formulate a series of environmental rights within a context of responsibility of the interdependent community of all life.[199]

Created through a participatory "global dialogue" amongst members of the global ethics movement, organisations, and the "world's great religions", the Charter Preamble proposed an ethical foundation for the emerging world community and a guide for ethical action at the individual and institutional levels (MacGregor, 2004, p. 85).

The Charter appeals to the common good for humanity while apparently avoiding an anthropocentric focus: "Recognise that peace is the wholeness created by right relationships with oneself, other persons, other cultures, other life, Earth, and the larger whole of which all are a part" (MacGregor, 2004, p. 85).

The Charter considers human rights as the basis of and the limitation to, human welfare and existence, focusing on the unity of human and non-human life. Some extracts illustrate this clearly: "We must join together to bring forth a sustainable global society founded on respect for nature, universal human rights, economic justice, and a culture of peace [...] The spirit of human solidarity and kinship with all life is strengthened when we live with reverence for the mystery of being, gratitude for the gift of life, and humility regarding the human place in nature [...] We affirm the following interdependent principles for a sustainable way of life as a common standard" (Bosselmann, 2008, p. 143).

The Earth Charter includes the recognition of the role of individuals and a kind of ecological citizenship and ecological duties granted to them.

2.2.1.1.1 Ecological Citizenship in the Charter

The Earth Charter appears to speak to citizens of the Planet rather than to the States or NGOs, using the word "we" in order to create the sense of common situation. And by using this universal "we" it proposes that "we" "are all in this (Planet) together"(MacGregor, 2004, p. 91).

The treaty introduces the non-territorial character of ecological citizenship (Norton, 2000, p. 1029): the political framework needed to develop

---

199 The Draft Earth Charter is published in Pacific Institute of Resource Management, 1992, Commitment for the Future: The Earth Charter and Treaties agreed to by the International NGOs and Social Movements, Paper Presented to the International NGO and Social Movements Forum Conference, Wellington, New Zealand, 11 June, the first principle states: "We agree to respect, encourage, protect and restore Earth's ecosystems to ensure biological and cultural diversity".

environmental protection strategies on various scales, local, regional, national and international (Dwivedi, 2006, p. 160; Markey, 2004, p. 76). This element of ecological citizenship is visible in the following passage of the preamble: "We are at once citizens of different nations and of one world in which the global and local are linked. [W]e must recognise that in the midst of a magnificent diversity of cultures and life forms we are one human family and one Earth community with a common destiny" (MacGregor, 2004, p. 85).

The ecological duties approach appears in the Earth Charter as well when duties *vis-à-vis* future generations and nature are announced in the Preamble: "Everyone shares responsibility for the present and future well-being of the human family and the larger living world"; and in fact, "with increased freedom, knowledge, and power comes increased responsibility to promote the common good".[200]

The Charter views human rights as vital to the fulfilment of human wealth, and recognises the role their exercise can play in reaching environmental aims. However, at the same time it acknowledges that the exercise of certain rights needs to be restricted through the exercise of ecological duties and responsibilities to secure the "long-term flourishing of Earth's human and ecological communities".[201]

In addition, the Earth Charter is set apart from other international agreements in that it "recognises that the successful achievement of ecological goals requires not only international commitment and legal regulation, but also basic changes in attitudes, values and behaviour of people" (Dwivedi, 2006, p. 160).

Some selected extracts will be illustrative here with regard to the recognition of ecological duties: Under the general heading of "Respect and care for the Community of Life" there are four broad duties: "1. Respect Earth and life in all its diversity. 2. Care for the community of life with understanding, compassion, and love. 3. Build democratic societies that are just, participatory, sustainable, and peaceful. 4. Securing the Earth's bounty and beauty for present and future generations".

---

200 Earth Charter Commission, 2002, Principle I:2b. MacGregor, 2004, p. 90.

201 Principle 4b of the Charter. NGOs drafted their own Earth Charter when UNCED was seen to fail in its objective. The NGO Earth Charter does not shy away from the task of accepting responsibility for nature and defines it in ecocentric terms. The Preamble states ,We accept a shared responsibility to protect and restore Earth and to allow wise and equitable use of resources so as to achieve an ecological balance and new social, economic and spiritual values.' The Draft Earth Charter is published in Pacific Institute of Resource Management, 1992, Commitment for the Future: The Earth Charter and Treaties agreed to by the International NGOs and Social Movements, Paper Presented to the International NGO and Social Movements Forum Conference, Wellington, New Zealand, 11 June, the first principle states: "We agree to respect, encourage, protect and restore Earth's ecosystems to ensure biological and cultural diversity".

In order to fulfil these four wide commitments, it is necessary to satisfy the following duties: "5. Protect and restore the integrity of Earth's ecological systems, with special concern for biological diversity and the natural processes that sustain life. 6. Prevent harm as the best method of environmental protection and, when knowledge is limited, apply a precautionary approach. 7. Adopt patterns of production, consumption, and reproduction that safeguard Earth's regenerative capacities, human rights, and community well-being. 8. Advance the study of ecological sustainability and promote the open exchange and wide *application of the knowledge acquired"* (Fowles, 2002).

In conclusion, there are many positive aspects to be noted with regard to the Earth Charter. Macgregor has argued that "Although difficult to measure, it is possible that it will contribute to a sense of common duties and purpose that can unite diverse social groups around the world". It may push to set up a global environmental citizenship and it has "undoubtedly brought the concerns of hitherto excluded and marginalised people to the international policy table. The individual is encouraged to think and act together as responsible earth citizens rather than to think independently or to be inquisitive about cultural, ethical, and political differences" (MacGregor, 2004, p. 91).

This ecological discourse developed in the Charter has been forgotten for a few years because unfortunately the responsibility to protect and improve the environment is still not considered to have the same relevance as environmental rights.[202]

Nevertheless, as mentioned above, in 2007 the French Head of State and senior UN representatives adopted the Paris Appeal calling for the adoption of a "Universal Declaration of environmental Rights and Duties" (Lador, 2010). The motivation behind this Appeal is the concern for environmental ethics leading to duties to complement any rights" (Bosselmann, 1993, p. 83).

---

202 It worth remarking that recent efforts have sought to provide more detailed lists of human duties. A group of former heads of states, joined in the InterAction Council, proposed such a text, to be adopted on the 50th Anniversary of the adoption of the Universal Declaration for Human Rights. UNESCO also has a text under consideration, drafted by a meeting of philosophers held in March 1997. The responsibilities it discusses are little more than an extension of human rights obligations to individuals and other non-state actors. Rather than limiting or 'balancing' the Universal Declaration of Human Rights with a declaration of responsibilities – which could provide a pretext for the state to limit existing rights – it would perhaps be better to attempt to extend the possibility of claiming human rights against non-state entities as well as against state actors". See Shelton, 2001, p. 185.

It can be concluded that ecological duties, while not yet contained in a legally binding instrument, are slowly gaining international recognition and influencing the growth of declaratory and also constitutional legal duties.[203]

Some authors have called to improve this aspect using criminal responsibility as an instrument to enforce ecological duties: "The importance of criminal responsibility is that it provides added incentives to refrain from harmful conduct by emphasising its culpable character, and in many cases, by allowing more stringent enforcement measures or penalties to be imposed" (Birnie, Boyle, & Redgwell, 2009, p. 329).

Unfortunately, the development of environmental liability law, as has occurred in many international agreements and declarations, remains largely an aspiration at the international level. Nevertheless, at a regional level, the most important example comes from the European Union where environmental liability has been recognised and drafted in a legally binding manner (Reiners, 2009).

2.2.2 Ecological Duties and Environmental Citizenship in the Aarhus Convention

The Aarhus Convention is "the first international legal instrument to provide a set of legal obligations", according to the Implementation Guide, the duties are principally *vis-à-vis* the future generations than *vis-à-vis* the Environment itself (Stec, & Casey-Lefkowitz, 2000, p. 29).

This can also be seen in its Preamble, where participation rights are extended to environmental citizens, and it is acknowledged that the ecological citizens have an obligation to future generations. In other words, the impact of present activities on the well-being of future generations must be taken into consideration (Stec, & Casey-Lefkowitz, 2000, p. 29).

State signatories to the Convention intended to encourage, by undertaking to guarantee a series of "citizens' rights in relation to the environment", what they affirmed in their ministerial declaration as "responsible environmental citizenship", to better enable all members of society to fulfil their duties. Moreover, the "engaged, critically aware public" is seen as both a player and a vital partner in the formulation and implementation of environmental policy (Pedersen, 2010; Koester, 2005, p. 31).

---

203 Moreover, a number of constitutions, including Article 51 A of the Indian Constitution, refer to the individual's duty to protect and improve the natural environment or some similar concept (see also Yemen Article 16, Papua New Guinea Article 5; Peru Article 123; Poland, Article 71, Sri Lanka, Article 28, Vanuatu Article 7, France).

The first preamble paragraph revokes Principle 1 of the Stockholm Declaration, which declares the "solemn responsibility" of man "to protect and improve the environment for present and future generations".

The preamble also recalls General Assembly Resolution 37/7 of 28 October 1982 in which there is a clear statement of the obligation of individuals to protect the environment, which accompanies the right of enjoyment of a healthy environment. Indeed, at paragraph 24 it states: "Each person has a duty to act in accordance with the provisions of the present Charter; acting individually, in association with others or through participation in the political process, each person shall strive to ensure that the objectives and requirements of the present Charter are met".

In addition, the fifth preamble Paragraph of the Aarhus Convention affirms: "the need to protect, preserve and improve the state of the environment and to ensure sustainable and environmentally sound development".[204]

In the context of the Treaty, the aforementioned statements establish that the procedural rights are not only important for the realisation of the substantive right to a healthy environment, but they also have a role to play in the fulfilment of ecological duties by helping to "protect, preserve and improve the state of the environment" (Stec, & Casey-Lefkowitz, 2000, p. 16).

This principle is better specified in the second part of the seventh preamble Paragraph, which states that: "every person has [...] the duty, both individually and in association with others, to protect and improve the environment for the benefit of present and future generations".

How the duties of protection, preservation and improvement affect the state of the environment is unclear. The first two words imply that environmental damage or degradation should be prevented while the verb "improve" appears to indicate that damage that has already been done should be repaired and the environment restored or renewed. The emphasis on "protection" and "renewal" rather than the "substitution" of man-made for natural resources means that "substitution" is not an option (Dobson, 1998, p. 45-46).

These paragraphs lay out the basis for the connection between public participation and basic human rights, the right to a healthy environment, as well as the duty to protect the environment for the benefit of present and future generations. In particular, the eighth paragraph declares: "Considering that, to

---

204 The term "sustainable development" has been used to embody an environmentally oriented approach towards economic development that meets the needs of the present generation without depriving future generations of the ability to meet their own needs. The definition found in the watershed Brundtland Report is "development that meets the needs of the present without compromising the ability of future generations to meet their own need".

be able to assert this right and observe this duty, citizens must have access to information, be entitled to participate in decision-making and have access to justice in environmental matters".

Although according to Waite "the convention represents the current high water mark of the law's attempt to meet Stone's demand for rights for the environment itself" (Waite, 2007, p. 410), it is clear that its approach is expressly anthropocentric. Indeed, the aim is to sustain an environment adequate to the health and well-being of every "living creature of present and future generations".

There is no suggestion that environment should be protected for the sake of other animals or for its own sake. According to Bell: insofar "as animals contribute to human health or well-being, we may have reason to protect their habitats. Insofar as other living and non-living entities contribute to human health or well-being, we may have reason to protect them. If something does not (and will not) contribute to human health and well-being, we have no reason to protect it" (Bell, 2004, p. 96).

Although the Convention holds on to the anthropocentric approach, in practice, there is also some kind of implicit eco-centric approach: there are no more doubts about the fact that human health and well-being are dependent on protecting the integrity of every constituent part of the global ecosystem.

For this reason some authors have argued that the Convention has a poly-centric approach. People may fulfil their duties under the procedural rights of the Convention, including bringing proceedings in order to protect their own individual interests, those of present and future generations, or those of the environment in its own right, even though this is not the same as rights for the environment, as advocated by Stone, or not even a truly eco-centric approach. The Convention is limited to "afford the possibility of humans exercising their rights to protect the environment" (Waite, 2007, p. 410).

Conclusion of Section I

It is possible to affirm that the old-fashioned view that international law is only concerned with the rights and obligations of states is open to objections when applied to the protection of the environment and it fails to fully represent the reality of the international legal system. Therefore, there are arguments for a more broadly-based system, which accords rights, or, in some cases, obligations to individuals.

The Aarhus Convention testifies the increased opening of international mechanisms with regard to access to environmental matters. Hence, this Treaty moves towards the creation of a global environmental democracy, recognising and encompassing several elements which are the basis for the construction of this new democracy.

In a formal dimension, the Treaty confirms the role of democracy to achieve environmental goals and provides ways to introduce elements of participatory and deliberative democracy, as will be shown in more depth in the next Section.

On a spatial scale, the Convention focuses on global and local levels. Its aim is to avoid the problem of democratic deficit, democratising not only the national or sub-national level, but also supra-national and international decision-making processes, which are playing an increasingly significant role in environmental governance.

Moreover, the Convention can be viewed as a step towards the creation of an environmental and ecological citizenship, which is detached from national boundaries and encompasses basic rights of the environmental citizen as well as basic duties of the ecological citizen.

Concerning environmental rights, the Aarhus Convention has also been identified as a measure of realising links between environmental protection and human rights norms, and it represents a contribution to a substantive human right to the environment. Nevertheless, the implementation guide to the Aarhus Convention argues that although the wording of the Convention is the clearest statement to date in international law pointing towards a human right to the environment, the Aarhus Convention does not explicitly state that any right exists *per se* (Stec, & Casey-Lefkowitz, 2000, p. 29).[205]

The use of procedural rights can be seen as a more effective way of achieving a healthy environment as well as a way of drawing attention to the creation of a substantive right (Soveroski, 2007, p. 269). This way, "the procedural rights set forth in the convention will contribute to the objective of an adequate environment for every person, which, in itself, adds an extra layer to the status of a substantive human right to a healthy environment" (Pedersen, 2010).

In fact, the Aarhus Convention's primary focus on procedural rights and its strong emphasis on non-state involvement represent the real *fait accompli* of the Aarhus Convention (Pedersen, 2010). It is important to emphasise that the procedural rights are not trivial,[206] but they offer a "genuine opportunity for real participation" (Bell, 2004b, p. 99).

Concerning ecological duties, the Aarhus Convention represents a giant step forward in the quest of strengthening citizens' environmental rights and duties, and with regard to the matching of rights with duties (Holder, & Lee, 2007, p.

---

205 The Implementation Guide also notes, however, that once the Aarhus Convention comes into force, it will represent a *fait accompli*, although the exact contents of a right will still be up for debate.

206 Most States have already amended or will have to amend their legislation and practice to meet the demands of Aarhus. See, for example, Tuesen, & Simonsen, 2000, p. 299, on Denmark, which is one of the most progressive countries in this area.

100). The Convention has determined that these procedural rights are essential for the achievement both of the right to a healthy environment, and also, no less importantly, of the possibility of individuals to fulfil their direct responsibilities towards others, including future generations and indirect responsibilities *vis-à-vis* nature.

In this respect, the rights and duties contained in the Aarhus Convention have the potential of actively improving environmentally responsible individual decisions and of becoming a tool by which environmental safeguards might be enhanced (Holder, & Lee, 2007)

Section II: Procedural Environmental Rights in the Aarhus Convention

This section will deal with the three official pillars, access to information, participation and access to justice, which are recognised by the Aarhus Convention and the two additional pillars, enforcement of environmental law and the Review of Compliance Mechanism, which are interpreted as quite revolutionary, and constitute tools which have never appeared in other international documents. The scope of this part is to show that the Aarhus pillars represent concrete examples of measures which could help to implement the theoretical model of environmental democracy at the international level, in Europe and beyond.

1 The Three Pillars Approach in the Aarhus Convention

1.1 First Pillar: Access to Environmental Information

Informational rights have been granted by environmental law, *via* soft law as well as through binding instruments at international and regional level.[207] Principle

---

207 Many international instruments focus on the right to information. It can be found in the United Nations Framework Convention on Climate Change. Its Article 6 provides that its parties „shall [...] promote and facilitate at the national and, as appropriate, sub-regional and regional levels, and in accordance with national laws and regulations, and within their respective capacities [...] public access to information [and] [...] public participation". Framework Convention on Climate Change, Article 6, May 9, 1992, 31 I.L.M. 849. Another example is the United Nations drawn up Draft Declaration of Principles on Human Rights and the Environment, which states: All persons have the right to information concerning the environment. This includes information, howsoever compiled, on actions and courses of conduct that may affect the environment and information necessary to enable effective public participation in environmental decision-making. The information shall be timely, clear, understandable and available without undue financial burden to the applicant. Cramer, 2009, p. 85.
See also e.g. Convention for the Protection of the Marine Environment of the North-East Atlantic, Article 9, Sept. 22, 1992, 32 I.L.M. 1069; Convention on Civil Liability for Damage Resulting from Activities Dangerous to the Environment, arts. 13-16, June 21, 1993, 32 I.L.M. 1228; North American Agreement on Environmental Cooperation, Article 2(1)(a),

10 declares that each individual shall have appropriate access to information concerning the environment, including information on hazardous materials and activities in their communities (Soveroski, 2007, p. 261). The theoretical importance of this right in combating environmental problems has already been analysed in Chapter I and at the international level it can be added that of the communications received by the Special Rapporteur on human rights and the environment referred to the failure to provide information (Shelton, 2001, p.185).

Access to information, the first pillar of the Convention, is the essential starting point for any public involvement in decisions (Lee, & Abbot, 2003, p. 88). The aim of the pillar guarantees that members of the public are conscious of what is occurring in their adjacent environment and moreover aims at the fact that the public shall be competent to participate in an informed manner.

The information pillar includes both the "passive" aspect, provided by Article 4 on access to environmental information; that is, the obligation on public authorities to respond to public requests for information, and the "active" aspect, provided by Article 5 on the collection and dissemination of environmental information. Article 3 reminds Parties that the Convention's provisions, including those in Articles 4 and 5, are minimum requirements and that Parties have the right to provide broader access to information for the public.

It should be remarked that these provisions are reinforced by the Kiev Protocol on Pollutant Release and Transfer Registers, commonly known as the PRTR Protocol, which was adopted in 2003, and entered into force on 8 October 2009.[208] The objective of the Protocol is to enhance public access to information

---

Sept. 14, 1993, 32 I.L.M. 1480; International Convention to Combat Desertification in Those Countries Experiencing Serious Drought and/or Desertification, June 17, 1994, 33 I.L.M. 1328; Convention on Cooperation and Sustainable Use of the Danube River, Article 14, June 29, 1994, available at www.icpdr.org/ icpdr-pages/drpc.htm; Energy Charter Treaty, arts. 19(1)(1), 20, Dec. 17, 1994; 33 I.L.M. 360; Protocol Concerning Specially Protected Areas and Biological Diversity in the Mediterranean, Article 19, June 10, 1995, 1999 O.J. (L 322) 3; Rotterdam Convention on the Prior Informed Consent Procedure for Certain Hazardous Chemicals and Pesticides in International Trade, Article 15(2), Sept. 10, 1998, available at www.fco.gov.uk/Files/ kfile/CM%206119.pdf; Protocol on Water and Health to the 1992 Convention on the Protection and Use of Transboundary Watercourses and International Lakes, Article 5(i), June 17, 1999, available at www.euro.who.int/ Document/Pehehp/ProtocolWater.pdf; Cartagena Protocol on Biosafety to the Convention on Biological Diversity, Article 23, Jan. 29, 2000, 39 I.L.M. 1027; International Treaty on Plant Genetic Resources for Food and Agriculture, Article 17, Nov. 3, 2001, available at ftp://ftp.fao.org/ag/cgrfa/it/ITPGRe.pdf. Also, at the European level the Directive 90/313/EEC on the freedom of access to information on the environment provided the legal basis for access to environmental information in the EC countries and in other countries in the UN/ECE region since its adoption.

208 The PRTR Protocol was adopted at an extra-ordinary meeting of the Parties in May 2003, in Kiev, Ukraine. The European Community and 36 States have signed the Protocol.

through the establishment of coherent, integrated, nation-wide pollution release and transfer registers, to facilitate public participation in environmental decision-making, and to contribute to the prevention and reduction of pollution in the environment. This international instrument is important because it represents, on the one hand, the first legally binding instrument related to pollutant release and transfer registers[209] and, on the other hand, it allows States outside the United Nations Economic Commission for Europe to join (Marshall, 2006, p. 126).

1.1.1 Passive Access to Information: Article 4 of the Convention

Article 4 sets out a framework through which members of the public can access environmental information from public authorities; in certain situations private parties can do likewise.[210] Once a member of the public has requested

---

Luxembourg recently became the first State to ratify the Protocol. Further information is available on its website www.unece.org/ env/pp/prtr.htm. See also Comba, 2009: "Le Protocole est salué comme une avancée importante pour garantir l'accès effectif du public aux informations concernant les polluants environnementaux, ceci en tant qu'étape essentielle pour l'exercice des droits énoncés dans la Convention d'Aarhus (Article premier), pour faciliter la participation du public au processus décisionnel et contribuer à la prévention et à la réduction de la pollution de l'environnement (Article 1 du Protocole). Le préambule souligne le bon rapport coût– efficacité de ces outils. Les États se sont engagés pour que les registres facilitent au maximum l'accès à l'information par les moyens électroniques, le public en ayant un droit sans qu'il ait à faire valoir un intérêt particulier (Article 11). Sur la base de l'Article 14 les parties, y compris les organisations régionales, doivent assurer que les personnes aient la possibilité de recourir devant une instance judiciaire (ou un autre organe indépendant et impartial). L'Article 23 discipline le règlement des différends parmi les Parties. Il reprend l'Article 16 de la Convention, cette-ci ayant crée un mécanisme obligatoire de résolution des différends. Les annexes II de la Convention et IV du Protocole disciplinent la procédure par arbitrage".

209 Article 1, PRTR Protocol.

210 Each Party shall ensure that, subject to the following paragraphs of this Article, public authorities, in response to a request for environmental information, make such information available to the public, within the framework of national legislation, including, where requested and subject to subparagraph (b) below, copies of the actual documentation containing or comprising such information: (a) Without an interest having to be stated; (b) In the form requested unless: (i) It is reasonable for the public authority to make it available in another form, in which case reasons shall be given for making it available in that form; or (ii) The information is already publicly available in another form. 2. The environmental information referred to in paragraph 1 above shall be made available as soon as possible and at the latest within one month after the request has been submitted, unless the volume and the complexity of the information justify an extension of this period up to two months after the request. The applicant shall be informed of any extension and of the reasons justifying it.

information, Article 4 establishes criteria and procedures for obtaining or refusal of such (Stec, & Casey-Lefkowitz, 2000, p. 53).

Paragraph 1 includes the general obligation for public authorities to grant environmental information in response to a request. The right of access extends to any person, without him or her having to prove or even state an interest. A request can be any communication by a member of the public to a public authority soliciting environmental information (Stec, & Casey-Lefkowitz, 2000).

As already shown, all persons have the right of access to information and the Convention provides a general rule of freedom of access to information. As a result, requests cannot be rejected even in cases where the claimant does not hold an interest in the information. Public authorities shall not call for any requirements which entail that the claimant has to show his or her motivation for receiving the information or how he or she intends to use it. The Convention obliges public authorities to supply the information "as soon as possible". It subsequently establishes a maximum time limit of one month. Nevertheless, public authorities in certain circumstances have the possibility under the Convention to find that the "volume and complexity" of the information excuses an extension of the one-month time limit to two months. In such situations, the authority must inform the claimant of this extension, by giving the reasons therefore, as soon as possible and at the latest by the end of the first month (Stec, & Casey-Lefkowitz, 2000).

The Aarhus Convention includes exceptions, as do all international agreements, created *via* political negotiation. There are eight specific cases that any authority may use as justification to refute an applicant's request. These exceptions include matters of national defence, the protection of trade secrets, and the protection of personal data and judicial or law enforcement matters in progress (Cramer, 2009, p. 100). The employment of an exemption is controlled by the words of the Convention, in particular by the final paragraph of Article 4: "The aforementioned grounds for refusal shall be interpreted in a restrictive way, taking into account the public interest served by disclosure and taking into account whether the information requested relates to emissions into the environment".

It has been underlined that a "blanket approach" to exceptions would be beyond the "spirit" of the Convention, and there is an obligation to engage in some sort of "consideration of the pros and cons of disclosure and confidentially: exceptions to access are provided not for convenience, but to protect genuinely competing public interest" (Holder, & Lee, 2007, p. 104).

In the case of refusals the reasons for them are to be issued in writing where requested. A time limit applies as for the supply of information: one month from the date of the request, with a provision for extending this by a further month where the complexity of the information justifies this. Where a public authority does not hold the information requested, it should either direct the requester to another public authority which it believes might have the information, or transfer the request to that public authority and notify the requester of this.

Nevertheless, there is a limitation in this Article centred on public authorities, providing no right of access in respect of information held by private parties. Article 4 applies only to information held by public authorities, very important information held by industry or subject to the convention's commercial and industrial exception is not covered, although a Protocol on Pollutant Release and Transfer Registers adopted in 2003 will require the industry to collect and report information about pollution emissions which parties must then make publicly available. The convention requires its members to encourage those operators to keep the public informed. The limitation is however compounded by the changing nature of public responsibilities because in recent years, in some states formerly public functions and activities have passed out of "government hands" (Lee, 2005, p. 153).[211]

This provision poses a more difficult issue which focuses on the functions of access to environmental information: whether access to environmental information obligations should apply to these sorts of entities. The Aarhus Convention nevertheless tries to guarantee that privatisation does not "take public services or activities out of the realm of public involvement, information and participation" (Stec, & Casey-Lefkowitz, 2000, p. 32) including in the definition of the authorities.[212]

1.1.2 Active Access to Information: Article 5 of the Convention

Article 5 establishes the obligation of the government to collect and disseminate information (Stec, & Casey-Lefkowitz, 2000, p. 49). It includes an extensive variety of different categories of information that Parties should supply to members of the public. Usually, it comprises information such as emergency information, product information, pollutant release and transfer information, information about laws, policies and strategies, and information concerning methods of receipt of information.

---

211   See also: Roberts, 2001, p. 243.

212   Article 2 Paragraph 2 provides "*Public authority means: (a) Government at national, regional and other level; (b) Natural or legal persons performing public administrative functions under national law, including specific duties, activities or services in relation to the environment; (c) Any other natural or legal persons having public responsibilities or functions, or providing public services, in relation to the environment, under the control of a body or person falling within subparagraphs (a) or (b) above; (d) The institutions of any regional economic integration organisation referred to in Article 17 which is a Party to this Convention. This definition does not include bodies or institutions acting in a judicial or legislative capacity*". Lee, 2005, p. 153; Roberts, 2001, p. 243.

The Aarhus Convention obliges the States to establish internal processes to ensure the ample flow of all significant environmental information and in addition concentrates on the real implementation of procedures for collecting and distributing information related to any threat to human health or the environment.[213]

The above-mentioned requirements apply to such private bodies with which governmental agencies cooperate, although the provision requires only that the State must encourage them (Cramer, 2009, p. 98).

Moreover, the State shall also encourage any private person to inform the public with regard to all activities and products which could have an impact on the environment, within the framework of voluntary labelling or auditing schemes.

Parties are required to "progressively" make environmental information publicly available in electronic databases, which can easily be accessed through public telecommunications networks (Wates, 2005a, p. 2).

Finally, the Convention requires that all parties publish, every three to four years, a national report on the state of the environment (Cramer, 2009, p. 97-99).

---

213   Article 5(1)(c). The text of this provision is worth quoting in its entirety: In the event of any imminent threat to human health or the environment, whether caused by human activities or due to natural causes, all information which could enable the public to take measures to prevent or mitigate harm arising from the threat and is held by a public authority is disseminated immediately and without delay to members of the public who may be affected.

1.2 Second Pillar: Environmental Participation

1.2.1 Three Possibilities to Participate

Public participation in decision-making is the second pillar of the Convention and its importance is fundamental because it is the link between all three principal pillars: public participation cannot be effective without access to information, as provided under the first pillar, nor without the possibility of enforcement, through access to justice under the third pillar.[214] The Convention is in theory "novel' by 'go[ing] well beyond familiar techniques of consulting neighbours over sitting decisions" (Stec, & Casey-Lefkowitz, 2000, p. 92) to foresee public participation in decision making at three stages: with regard to "decisions on specific activities" as defined in Article 6; concerning plans, programs and policies relating to the environment, listed in Article 7; and finally, in relation to the preparation of executive regulations and/or generally applicable legally binding normative instruments as presented in Article 8 (Nadal, 2008, p. 37).

1.2.1.1 Public Participation in Decisions on Specific Activities: Article 6 of the Convention

Article 6 concerns participation in decisions permitting certain activities listed in Annex I of the Convention; for example, activities within chemical installations and waste management, or other activities which may have a significant effect on the environment. The emphasis here is not only specific and local but also 'reactive' and 'defensive'. The public has the opportunity to react to and defend themselves against proposals for activities with significant environmental impacts.

Activities under Article 6 generally include activities subjected to the environmental impact assessment (EIA) procedure under the UNECE Espoo Convention on environmental Impact Assessment in a Trans-boundary Context.

An activity originally omitted from the mandatory requirements of Article 6 was decision-making concerning genetically modified organisms. However, at their second meeting in May 2005, the Meeting of the Parties adopted an amendment imposing a public participation requirement in decisions on the

---

214 This right was already included in a large number of treaties: for example, see ICCPR, n. 7 above, Article 25; American Declaration, n. 32 above, Article 20; African Charter, n. 37 above, Article 13; American Convention, n. 34 above, Article 23; and the Convention on environmental Impact Assessment in a Trans- boundary Context (Espoo, 25 February 1991). Soveroski, 2007, p. 261.

deliberate release and placing on the market of genetically modified organisms. Participation should commence already early in the process, when options are still open, and moreover, due account must be taken of the result of public participation (Nadal, 2008, p. 38). In fulfilment of this, required elements embrace: public notice of the projected activity, complete information on the planned activity, transparent opportunities for public comment and participation, reasonable timeframes for participation, and "the public concerned shall be informed, either by public notice or individually as appropriate, early in an environmental decision-making procedure, and in an adequate, timely and effective manner" of a number of matters relating to the permit application.[215]

Paragraph 7 varies from the majority of the other provisions of Article 6 in that here the Convention accords rights not only to the public concerned, but also to the totality of the public. For this reason it has been considered as the "backbone of public participation" (Holder, & Lee, 2007, p. 101) and constitutes the only reference to deliberative process within the Convention.

Thus, any member of the public has the right to submit comments, information, analyses or opinions for the duration of the public participation procedures. The public authority cannot refuse any such comments, information, analyses or opinions by arguing that the specific individual was not a part of the public concerned.

However, it should be remarked that Article 9(2) which is the enforcement of Article 6 applies only to the "public concerned". Nevertheless, it seems to be the objective of the Convention that any member of the public, who by submitting comments in writing or at a hearing actually participates in a public participation procedure, acquires a new status: the status of a member of the public concerned. The Convention refers to two potential deliberative means for the presentation of comments, information, analyses or opinions, through written presentation, public hearings or enquiries with the claimant. The public hearings or enquiries allow the claimant to submit the project, and answer to questions and comments. Public hearings also offer a place for discussion among persons concerned by the decisions (Stec, & Casey-Lefkowitz, 2000, p. 93).

Furthermore, under Article 6(8), decision-makers are required to take *due account* of public participation regarding the environmental aspects of the projected activity. That means that "public views cannot be simply ignored,

---

215 The information to be provided includes: the proposed activity and the application on which a decision will be taken; the nature of possible decisions or the draft decision; the public authority responsible for making the decision; the envisaged procedure including opportunities for public participation; the fact that the activity is subject to a national or transboundary environmental impact assessment procedure.

however giving the public only a limited stake in the final decisions" (Lee, & Abbot, 2003, p. 94).

Article 6(9) strengthens the above mentioned obligation by requiring that their objections shall be specified for the final decision.[216] Hence, the obligation to take due account appears not only as a procedural requirement but also, as it has been noted, as "a substantive one to integrate environmental justice considerations into the final decision" (Lee, & Abbot, 2003, p. 94).

1.2.1.2 Public Participation Concerning Plans, Programmes and Policies Relating to the Environment: Article 7 of the Convention

This Article requires parties to make "appropriate practical and/or other provisions for the public to participate during the preparation of plans and programmes relating to the environment". Commentators have noted that the term "relating to the environment" is wide, "covering not just plans or programmes prepared by an environment ministry, but also sectoral plans such as transport, tourism, etc, where these have significant environmental implications" (Wates, 2005a, p.6).

Participation requirements related to plans and programs are not specified in similar detail as in the case of Article 6, because the strength of the participatory requirements diminishes as we move from plans and programmes, which are often regional, to policies and executive regulations, which can also be national. Public participation should take place in a transparent and fair framework and also follow numerous principles which are provided under Article 6, as well as realistic timeframes, early participation, and due attention to the result of the participation.

Article 7 devotes only one sentence to policies: "To the extent appropriate, each Party shall endeavour to provide opportunities for the public participation in the preparation of policies relating to the environment". The institution of representative democracy is required to be consulted only "to the extent appropriate" and one has no obligation to take "due account" of any public comments (Bell, 2004, p. 99).

---

216 Developments under the Human Rights Act will require reasons to be given for any decision in at least planning law, and Government seems amenable to introducing such a change in its ongoing reform of the planning system. A duty to give reasons is however fundamentally linked with review of the decision, on which there is less progress – giving reasons enables such review and is most effective if review is possible. Lee, & Abbot, 2003, p. 94.

Article 7 differentiates between plans and programs on the one hand, and policies on the other. As far as the former are concerned, the provision includes elements of Article 6, especially relating to the time-frames and occasions for public participation, as well as the commitment to guarantee that public participation is taken into consideration. With respect to the regulation of policies, there is no express incorporation of any of the principles of Article 6.

The Implementation Guide of the Convention has suggested cohesion with strategic environmental assessment (SEA) as a method of implementing Article 7 through public participation procedures (Stec, & Casey-Lefkowitz, 2000, p. 113-114). The obligation that States guarantee that "due account is taken of the outcome of public participation" means that "there must be a legal basis to take environmental considerations into account, in plans, programmers and policies" (Stec, & Casey-Lefkowitz, 2000, p. 113-114).

Through the access to justice provisions of Article 9, paragraph 2 the members of the public can enforce the obligations under Article 7, which requires that Parties take legislative steps to adopt guarantees for the rights contained in this Article. If States previously have established guarantees, these must be preserved. Furthermore, in those instances where they do not have guarantees and do not implement new guarantees, prospects for the enforcement of commitments, according to the Implementation Guide: "must be based on Article 9, paragraph 3, which provides for the right of citizens to bring actions in cases of violations of environmental law" (Stec, & Casey-Lefkowitz, 2000, p. 115).

1.2.1.3 Public Participation During the Preparation of Executive Regulations and/or Generally Applicable Legally Binding Normative Instruments: Article 8 of the Convention

Article 8 of the Convention addresses public participation in the preparation of executive regulations and legally binding normative instruments. Article 8 requires only that the result of public participation shall be taken into account as far as possible in the preparation of executive regulations and legally binding normative instruments. Policies, executive regulations and law provide the context in which plans, programmes and decisions about specific activities aremade.

Although the Treaty does not apply "to bodies acting in a legislative capacity", this Article "would apply to the executive stage of preparing rules and regulations even if they are later to be adopted by parliament" (Lee, & Abbot, 2003, p. 94).

This provision is "quite novel" (Lee, & Abbot, 2003, p. 101), since it not only concerns individual decisions, or decisions by independent agencies, but also legislative decisions. It stipulates that draft rules be published or otherwise be made publicly available, that the public should be granted the right to express criticism and comments directly, or through representative consultative bodies, and that the outcomes shall be considered.

Moreover, this provision expands well beyond classic pollution or conservation law, and could include, for instance, decisions on energy or transport.

Article 8's wording is somewhat weaker than Article 6 and is even less precise than Article 7, and a justification for this with regard to the applicability of the Convention to law-making can be found when analysing the drafting procedure. Here it can be seen that this issue was carefully debated during the whole process of negotiations. Nevertheless, governments did not agree to provide detailed requirements for parliaments, since they considered this to be a prerogative of the legislative body (Stec, & Casey-Lefkowitz, 2000, p. 122).

Nevertheless, this has not stopped the Aarhus Compliance Committee from finding at least one State in breach of the Convention.[217]

Despite the vague character of the provision, it can be viewed as a possibility given to States to interpret the provisions in a different way and it could be a noteworthy political instrument in the incorporation of environmental preoccupation into other policy matters (Lee, & Abbot, 2003, p. 101).

1.3 Third pillar: Access to Justice

Several international human rights instruments guarantee a fair trial, in other words a right to equal access before courts or other independent and impartial tribunals.[218] The starting point of access to justice in environmental matters is Principle 10 of the Rio Declaration which stipulates "effective access to judicial and administrative proceedings, including redress and remedy, shall be provided".

This provision has been implemented by the Aarhus Convention, which is a unique Convention setting out minimum standards of access to legal review procedures.[219]

---

217   Dalma Orchards: Armenia, Compliance Committee, 2006.

218   UDHR Article 10; International Covenant on Civil and Political Rights, Article 14; European Convention on Human Rights and Fundamental Freedoms, Article 6. Ebbesson, 2009.

219   There are also other regional agreements which make also broad provisions for environmental claims: the 1993 North American Agreement on Environmental Cooperation is another example. "Article 6 gives persons with a ,legally recognised interest' the right to bring proceedings to enforce national environmental laws and to seek remedies for environmental harm; Article 7 provides for these proceedings to be fair, open and equitable and to conform to standards of due process. One unusual provision of this agreement allows individuals and NGOs to complain to the secretariat that a state's party is failing to

Before the Aarhus Convention, most national and international law systems featured laws generally allowing only the individual to use the justice system to seek a remedy for his grievance; those who were not personally affected were "unable to go before courts as proxies for the victim or aggrieved party". Hence, if there was no personally affected individual at all, as a general rule, there would be nobody to remedy certain actions, even if these actions were in violation of a law" (Schall, 2008, p. 417). But now, in accordance with Article 9, the access to justice pillar not only underpins the first two pillars but also "points the way to empowering citizens and NGOs to assist in the enforcement of the law" and also helps to overcome difficulties "such as the non-transposition into domestic law of international treaty obligations which are of a non-self-executing character" (Redgwell, 2007, p. 159).

1.3.1 Article 9 of the Convention

Article 9 contains two categories of provisions which should be analysed separately since they are completely different.

First, access to justice means that members of the public have legal mechanisms, that could be used against potential violations of the two other pillars, – access to information and public participation.

Second, access to justice means that the public is equipped with legal mechanisms which they can use to gain review of potential violations of domestic environmental law and thus, the public's ability to help enforce environmental law is acknowledged. This third part of Article 9 is not linked with the other pillars of the Convention, but it should be considered a new right recognised by the Convention.

Hence, paragraphs 1 and 2 are directly related to the internal provisions of the Convention while paragraph 3 reinforces external domestic standards. The specificity of this form of "external review" (Redgwell, 2007, p. 168) has led to it being considered a fourth pillar (Jóhannsdóttir, 2008, p. 221) of the Convention. This is also due to the fact that it has no connection with either of the first two pillars of the Convention.

For this reason, it will be studied later separately from the normal access to justice.

---

enforce its environmental legislation". See: Birnie, Boyle, & Redgwell, 2009, p. 291.

1.3.2 Access to Justice to Enforce the Two Pillars

The first two paragraphs of Article 9 are related to the first two pillars of the Convention.

First, Article 9(1) acknowledges that any person, who believes that his or her request for information was ignored, wrongfully refused, or inadequately answered, has, in accordance with national law, access to a judicial or non-judicial review procedure.[220]

Under this provision, any person has a right to exercise the review procedures and has standing to challenge decisions made under Article 4. Moreover, Article 9(1) is in conformity with Article 4's language, which grants any member of the public the right to request information. In addition, this paragraph provides that the review procedure must be before a court of law or any other "independent and impartial body established by law".

The significance of "independent and impartial body" can be explained by the Convention for the Protection of Human Rights and Fundamental Freedoms: independent and impartial bodies do not have to be courts, but must be "quasi-judicial, with safeguards to guarantee due process, independent of influence by any branch of government and unconnected to any private entity".[221] States have the obligation to guarantee that the public has access to faster and less expensive review procedures than reviews in courts (Stec, & Casey-Lefkowitz, 2000, p. 127). Moreover, the public authority has to be bound by final decisions.

Second, paragraph 2 provides that members of the public and any NGOs which have *sufficient interest* or who maintain "impairment of a right where the administrative procedural law of a Party requires this as a precondition" are able to "challenge the substantive or procedural legality of any decision, act or omission" under Article 6, and also any decision under other relevant provisions of the Convention (Kravchenko, Skrylnikov, & Bonine, 2003, p. 27). The general provisions of Article 3 and the provisions concerning the collection

---

220   "Each Party shall, within the framework of its national legislation, ensure that any person who considers that his or her request for information under Article 4 has been ignored, wrongfully refused, whether in part or in full, inadequately answered, or otherwise not dealt with in accordance with the provisions of that Article, has access to a review procedure before a court of law or another independent and impartial body established by law. In the circumstances where a Party provides for such a review by a court of law, it shall ensure that such a person also has access to an expeditious procedure established by law that is free of charge or inexpensive for reconsideration by a public authority or review by an independent and impartial body other than a court of law".

221   Some countries have chosen to create a special, independent and impartial body to review access-to-information cases. For example, in 1978 France established the Commission for Access to Administrative Documents (CADA). Stec, & Casey-Lefkowitz, 2000, p. 126.

and dissemination of information in Article 5 could be provisions that would fall under the expression "other relevant provisions".

In determining the standing of the public concerned, the Convention defers to national law, but emphasis is given to the objective of giving the public concerned wide access to justice. Furthermore, bodies that fulfil the Convention's definition of the public concerned, which includes NGOs, are automatically considered to have a sufficient interest, or rights capable of being impaired (Lee, & Abbot, 2003, p. 101).

Art. 9's "sufficient interest" is not defined by the Aarhus Convention; however, it appears for some commentators to be narrower than the "public concerned" employed in Article 6, and the parties could not agree on how far it provides for public-interest litigation by NGOs (Birnie, Boyle, & Redgwell, 2009, p. 295). Aarhus "creates a fiction concerning standing requirements, as the necessary 'sufficient interest' to institute proceedings is already constituted by the interest of any NGO acknowledged by national law.[222] Therefore, there is a general objective of Aarhus to give the public concerned wide access to justice" (Schall, 2008, p. 417). Hence, persons or groups who meet these conditions will still need to satisfy the requirements of national law, but with the provision "that any such requirements must be consistent with the objective of giving the public concerned wide access to justice within the scope of Convention" (Redgwell, 2007, p. 169).

In its first ruling, the Compliance Committee held that although what constitutes a sufficient interest and impairment of a right shall be determined in accordance with national law, it must be decided with "the objective of giving the public concerned wide access to justice within the scope of the convention".[223] "The non-discriminatory application of rights of public participation and access to environmental justice under Article 3(9) will also include trans-boundary claimants, and may thus facilitate resolution of trans-boundary environmental disputes" (Birnie, Boyle, & Redgwell, 2009, p. 295).

2 Two More Pillars of Aarhus Convention?

This book has decided to call for different classifications of the following two rights granted to the public in the Convention which generally are analysed under the "umbrella" of the third pillar. The reason is that, despite this content

---

222 Aarhus Convention (n. 7), Art. 9(2).
223 Compliance Committee: Bond Beter Leefmilieu Vlaanderen VZW, 2006.

being near to the possibility of access to justice, both are original because never before were they granted to individuals.

2.1 Enforcement of Environmental Law: Article 9(3) of the Convention, the Fourth Pillar

Article 9(3) creates an additional category of cases, where citizens have access to administrative or judicial procedures to challenge acts and omissions, whether or not these are related to the information and public participation rights, by private persons and public authorities which contravene national law relating to the environment.

The eighteenth preamble paragraph as well as the Sofia Guidelines already provided standing to certain members of the public to enforce environmental law in a direct or indirect manner. Concerning direct citizen enforcement, citizens are given standing to go to court or other review bodies to implement the law rather than just to redress personal damage. Indirect citizen enforcement means that citizens can contribute to the enforcement process through, for instance, citizen complaints (Stec, & Casey-Lefkowitz, 2000, p. 130).

Moreover, the Convention allows a person to challenge acts and omissions by private persons and public authorities which contravene provisions of national law relating to the environment. This wording includes on the one hand, failures to take action provided by law, and on the other, actions that themselves infringe the law (Bonine, 2003, p. 31).

This provision obliges States to guarantee standing to enforce environmental law for those citizens who meet criteria provided for by national law (Stec, & Casey-Lefkowitz, 2000, p. 130). Standing under Article 9(3) is even more restrictive than under Article 9(2). The reason therefore is "the price paid for the right to challenge violations of national laws" relating to the environment or omissions by public authorities (Redgwell, 2007, p. 169).

National law must make the decision whether redress is administrative or judicial, and establish standing requirements in order to challenge acts or omissions in connection with national environmental law (Redgwell, 2007, p. 169). It should be remarked that judicial interpretation could play a significant role in the enforcement of the Aarhus Convention (Savoia, 2003, p. 39).

Concerning standing, it has been emphasised that several broad legislative models are possible: *actio popularis;* NGO standing; legal rights standing and sufficient interest standing. Concerning *actio popularis*, some countries use a model in which legislation declares that *any person* can sue the government when it breaks the law, an *actio popularis*. This is completely in accordance with Article 9 of the Aarhus Convention, even though it is not obligatory by the Convention.[224]

Under the second model, numerous States recognise a special right to NGOs to have standing rights concerning possible judicial action without demonstrating that they have a personal interest or are affected by a decision. National law has either to identify the features of NGOs that are given standing, or it requires that an authority will establish a register of NGOs that are automatically recognised standing and legalised to take to trial illegal acts by government to the courts (Bonine, 2003, p. 31).

Finally, sufficient interest standing grants legal standing to those who are *affected*. This usually may be granted to all persons, or be drafted as part of the legislation granting NGOs standing.

Article 3(4) of the Aarhus Convention declares that, unless there is national legislation which imposes special requirements, interest is simply the fact that an NGO is devoted to environmental protection (Bonine, 2003, p. 31).

---

224 The Netherlands may well have the least restrictive legislative criteria in Europe for accessing the courts. Furthermore, the Netherlands links administrative standing and judicial standing by allowing "anyone" to participate in the consultation process with a public authority and then granting anyone who has lodged objections at the consultation stage the right to ask a court for judicial review of the decision. Additionally, the Netherlands also extends standing to NGOs in civil lawsuits much like Italy, Switzerland, or many German *Lander*. See Bonine, 2003, p. 31-37. According to a 1992 study, Switzerland was the first country to legislate a right of action (or standing to sue) for environmental NGOs. In Switzerland, Article12 of the Federal Nature and Heritage Conservation Act of 1966 allows appeals against administrative decisions to the Supreme Court, for nationwide nature associations. The same can be found in Article 55 of the environmental Protection Act of 1983 for nationwide nature NGOs, provided they were founded at least ten years before the law suit and are officially recognised by the federal government. A third law, the Trails and Footpaths Act of 1987, also uses this accreditation procedure. In Italy, Articles 13(1) and 18(5) of Law no. 349 of 1986 grant environmental associations the right to sue in administrative courts if they have been recognised for this purpose in a ministerial decree.

In France, the prerequisites for standing if the plaintiff has an 'interest to act'are such that everyone, even environmental associations not recognised by French law, can bring a public interest action. "Therefore, the plaintiff has only to demonstrate his interest in the case by showing a 'grief moral'. NGOs have access to a court if they are able to demonstrate a connection between their objectives and activities on the one hand and the interests at stake on the other hand. [...] German law on access to justice and PIL, however, fundamentally differs from its English and French counterparts. Section 42(2) of the Administrative Courts Act applies the so-called 'protective norm doctrine' (Schutznormtheorie). Therefore, access to justice in Germany is based on the protection of individual rights. Locus standi is only conferred on the plaintiff if the rights that appear to be violated are intended to protect the plaintiff himself, as an individual, and not merely the general public interest. NGOs and individuals thus have to demonstrate a private interest in the case, showing the violation of a 'subjective' right construed for their protection. Official actions with adverse effects on the environment generally cannot be challenged in the public interest". Schall, 2008, p. 417.

Thus, Article 9(3) recognises the significance of the public enforcement of environmental law by providing for direct action against polluters or regulators and this suggests a continued "monitoring" type role for the public (Lee, & Abbot, 2003, p. 101).

Hence, this provision from an idealistic point of view is a quite big *revolution* in the field of environmental law enforcement and thus it might be considered a fourth pillar. Unfortunately, though, direct citizen enforcement, as a model of a *citizen suit* has still not been developed throughout Europe (Lee, & Abbot, 2003, p. 101).

The Task Force on Access to Justice (TFAJ) has underlined the development of guidelines for the identification of best practice to assist with implementation of this paragraph, but the diversity of national initiatives and a lack of political will has led to little progress in this issue.

Another difficulty with Article 9(3) is the absence of a treaty definition of national law relating to the environment, leaving it open to parties to define it in their implementing measures. The TFAJ has considered this issue, during the drafting process; this sentence "is broader than national legislative provisions specifically aimed at the protection of the environment and includes any provisions of national law, whether statutory or regulatory, whose enforcement has an effect on the state of the elements of the environment or on factors and activities or measures affecting or likely to affect these elements".[225]

However, no agreement could eventually be reached to insert such a provision in the draft decision which was submitted to the Meeting of the Parties for adoption; moreover, no agreement exists on the topic related to a national law wider than specific environmental legislation.[226] As a result, the decision, as adopted by the Parties, does not refer to this question at all. The notion remains subject to interpretation by Parties in accordance with their domestic legal systems. Nevertheless, there are strong reasons in favour of a broad interpretation as argued by Pallemaerts: "in view of the Convention's definition of environmental information and of the fact that the drafters of the Convention apparently deliberately chose not to use the more common concept of "environmental law", which may have a clear-cut, technical meaning in some legal systems" (Pallemaerts, 2009).

To conclude, this fourth pillar in the Aarhus Convention clarifies that it is not only the purpose of environmental authorities and public prosecutors to

---

225 Report of the second meeting of the Task Force on Access to Justice, UN Doc. MP.PP/WG.1/2004/3, 8 January 2004, *available at www.unece.org/env/documents/2004/pp/mp.pp/wg.1/mp.pp. wg.1.2004.3.e.pdf,* Annex, p. 15, para. 17.

226 Report of First Meeting, para. 28. See also Redgwell, 2007, p. 170.

enforce environmental law, but that the public plays a role as well, to fulfil the environmental duty to conserve and protect the environment for future generations (Stec, & Casey-Lefkowitz, 2000, p. 130-131).

2.2 Review of Compliance: Article 15, a Fifth Pillar

Remarkable progress has been made by the Aarhus Convention in advancing access to justice with regard to individual rights and as a tool for the implementation of environmental law, but the doctrine has underlined some limits and obstacles on the road to the effective realisation of the right to environmental justice.[227] In particular, the Aarhus Convention has been criticised because it ensures access to justice only at the level of domestic law and it does not provide a supranational forum to "adjudicate claims concerning the availability of procedural guarantees to challenge environmental wrongs" (Francioni, 2008, p. 31).

However, the Convention is evolving and the decision to establish a non-compliance procedure before a committee of independent experts marks a positive step toward "the setting up of a review mechanism for possible violations by States parties for their obligation to guarantee appropriate remedial action to all persons subject to their jurisdiction who may lament environmental injuries or abuses" (Francioni, 2008, p. 25).

Thus, the innovative element of the Convention's institutional mechanism is the Compliance Committee (Pallemaerts, & Moreau, 2004), established by Article 15, because it "represents an important and inventive approach to the supervision of international agreements" (Pedersen, 2010). Furthermore, his Article is especially important in the light of the absence of supranational forums for the direct enforcement of international environmental law.

Review of compliance is such an important tool not only because it is a way to assure access to justice—and not just at the domestic level—but also because the role of the public is stressed.

Such provisions granting to individual citizens and NGOs the right to actually participate in the monitoring, by an international body, of state compliance

---

227 First, a limit of the Aarhus Convention is that environmental justice "may remain in the domestic law as a result of technical rules on *locus standi* of potential complainants and of available remedies to prevent or redress environmental violations. As far as locus standi is concerned, restrictive domestic legislation may impede the admissibility of public interest intervention and the bringing by plaintiffs of an *action popularis*, especially when such actions are not related to the reparation of damage already occurred but rather to forestalling of prospective damage that is feared to follow the realization of a project".. See for this topic: Francioni, 2008, p. 31.

with legal obligations is "unprecedented in international environmental law" (Pallemaerts, 2004, p. 20).

In fact, for the first time in international environmental law, provisions contained in a Convention open up the possibility of the establishment of a review mechanism accessible not only to states, but also to individuals.[228]

The most innovative part of Article 15 provides for the establishment of "arrangements of a non-confrontational, non-judicial and consultative nature",[229] and for reviewing compliance of parties which "shall allow for appropriate public involvement and may include the option of consideration of communications from members of the public on matters related to this Convention".

It should be observed that the phrase "matters related to this Convention" as used in this provision, is rather open-ended and could be interpreted "not only as referring to violations of the specific procedural rights, guaranteed by the Convention, but also conceivably as not precluding communications about the observance of other substantive rights to healthy environment recognised as an objective in Article 1".[230]

In order to achieve this, Decision I/7, taken by the Meeting of the Parties at their first meeting in October 2002, created the Aarhus Convention Compliance Committee.[231] The annex to Decision I/7 establishes the organisation and tasks of the Compliance Committee as well as the processes of the review of compliance. The compliance mechanism is unique in a number of aspects.[232]

---

228 To date, four multilateral environmental agreements have compliance regimes in operation, including the Montreal Protocol on Substances that Deplete the Ozone Layer, 1987, the Convention on Long-range Transboundary Air Pollution, 1979, the Convention on International Trade in Endangered Species of Wild Fauna and Flora, Washington 1973, and the Convention on the Conservation of European Wildlife and Natural Habitats, Bern 1979.

229 This phrase has several implications. The first is that the intention of compliance review is not to point the finger at Parties that are in violation of the Convention, but to recognize and assess the shortcomings of Parties and to work in a constructive atmosphere to assist them in complying.

230 Pallemaerts, 2002, p. 17.

231 ECOSOC, ECE, Meeting of the Signatories to the Convention on Access to Information, Public Participation in Decision-making and Access to Justice in environmental Matters, *Annex to the Addendum to the Report of the First Meeting of the Parties: Decision I/7 Review of Compliance* 4, U.N. Doc. ECE/MP.PP/2/Add.8 (Apr. 2, 2004), available at *available at www. unece.org/env/pp/documents/mop1/ece.mp.pp.2.add.8.e.pdf.*

232 The Committee has three main functions: (a) To consider submissions by Parties, referrals by the secretariat, and communications from the public; (b) To prepare a report on compliance with or implementation of the Convention for the Meeting of the Parties; and (c) To monitor, assess and facilitate implementation of and compliance with each Party's obligation to regularly report on their implementation of the Convention. In addition, the Committee may examine compliance issues and make recommendations on its own initiative. Marshall, 2006, p. 127.

Thus, the compliance mechanism provides that communications concerning a Party, for whom the Convention has entered into force, may be brought before the Committee by one or more members of the public concerning that Party's compliance with the Convention. Such communications from the public are authorised to be submitted in written or electronic form, and must be complemented by "corroborating information".[233]

The Committee may refuse to consider such a communication which it believes to constitute an "abuse of the right to make such communications" or "manifestly unreasonable".[234] Moreover, the Committee must determine whether domestic remedies were accessible and whether they were employed. However, the expression "absolute exhaustion of domestic remedies" has not been clarified by the Decision I/7.[235]

It has been noted that "the public's right to participate in the Committee's processes reflects the concept of participation enshrined in the Convention itself" and the mechanism is also based on the deliberative theory. In particular, for the most part, Committee meetings are open to the public and the public may participate in hearings and discussions of particular cases as observers.[236]

Another innovative aspect is that Committee members are nominated by State parties, signatories and NGOs, and elected by the Meetings of the Parties by *consensus* or, failing *consensus*, by secret ballot; and they serve in their personal capacity.[237]

It has to be noted that some legal scholars and NGOs have criticised these review systems for lacking teeth (Ebbesson, 2007, p. 683). Nevertheless, the Convention goes well beyond other international environmental arrangements in providing access to a review procedure for members of the public and this opening "to public participation by civil society has already produced remarkable results in the functioning of the Committee".[238]

---

233   Idem, Annex 19.

234   Idem, Annex 20.

235   Idem, Annex 21. Kravchenko, 2007, p. 1.

236   The Committee will, however, hold closed Sessions of decision-making, such as when deliberating on findings, measures and recommendations and if necessary to protect the confidentiality of information under the grounds discussed above. Marshall, 2006, p. 134-135.

237   For an NGO perspective on the compliance committee, see Guidance Document on Aarhus Convention Compliance Mechanism (undated), available at *www.unece.org/env/pp/ compliance/ manualv2. document>*. Morgera, 2005, p. 140.

238   As of late 2008, the Aarhus Compliance Committee, which arbitrates alleged violations of the treaty, has reviewed twenty-three allegations of nondisclosure of environmental information by a party to the treaty. Thus far, four of those cases (two brought by citizens in Kazakhstan, and one each by citizens in Hungary and Ukraine) have forced the Aarhus Compliance Committee to make authoritative judgments on the treaty's

Moreover, according to Fitzmaurice the compliance committee has a significant role in promoting environmental justice and even in contributing to the implementation of democratic governance, and the complaint procedure has been seen as "part of the fabric of the new world order, through expanding public participation of civil society" (Fitzmaurice, 2009a, p. 211).

Therefore, one may suggest that it could be seen also as a fifth pillar of the Convention towards the construction of global environmental democracy in which the validity and legitimacy of norms directly depends on the participation of citizens in their formation and application, as is also relevant in a global community (Fitzmaurice, 2009a).

Conclusion of Section II

In conclusion, it is possible to confirm that the Aarhus Convention is, for the moment, "the most ambitious venture in the area of environmental democracy"[239] within international environmental law. The major results of the Aarhus Convention would be the use of procedural avenues as a way of achieving substantive outcomes and its strong focus on the empowerment of individuals and NGOs.[240]

In addition, the Compliance Committee to the convention is likely to further improve notions of transparency and openness while at the same time being a novel institution in itself and thus it can be seen as the fifth pillar of the Convention (Pedersen, 2010).

Despite the above-mentioned important outcomes, this Treaty has received criticism for its weaknesses and shortfalls (Brooke, 2006, p. 341). The right to participation and the right to access to justice are not as far-reaching as a first glance might indicate (Fitzmaurice, 2003, p. 341; Rose-Ackerman, & Halpaap,

---

fundamental matters of access to information and public participation. Cramer, 2009, p. 100-101; Kravchenko, 2007, p. 1.

239   UNECE, Introducing the Aarhus Convention, available at www.unece.org/env/pp/welcome.html (quoting Kofi A.).

240   Kravchenko, 2007, p. 2. The author noted that the system of the Compliance Committee has inspired the Parties to the Protocol on Water and Health to the 1992 Convention on the Protection and Use of the Transboundary Watercourses and International Lakes to adopt an almost identical system of compliance.
See also U.N. Econ. & Soc. Council [ECOSOC], Econ. Commission' for Europe, Report of the Meeting of the Parties to the Protocol on Water and Health to the Convention on the Protection and Use of Transboundary Watercourses and International Lakes, U.N. Doc. ECE/MP.WH/2/Add.3, available at www.unece.org/env/documents/2007/wat/wh/ece.mp.wh.2_add_3.e.pdf.

2001). In fact, it must be pointed out that several problems and ambiguities in the Aarhus Convention have been identified by the legal doctrine. It is therefore considered a "fairly weak legal document, given its quite vague and permissive character and the absence of adequate enforcement mechanisms" (Lee, & Abbot, 2003, p. 81).

However, the ratification process by the States of the Aarhus Convention solves some of these difficulties.[241] Indeed, it is well-known that for international environmental law to be effective, it relies upon domestic orders for its implementation as well as its enforcement. Furthermore, the elements of environmental democracy have to be reflected also at the local level through regional and national regulation. Thus, some of the obligations within the Aarhus Convention which are considered weak are likely to be given some real teeth *via* regional and national legislation (Wates, 2005b, p. 393).

Hence, in conclusion it can be stated that the Convention makes a potentially powerful proclamation with regard to the significance of public participation in an ample variety of decisions and it should therefore not be forgotten that the treaty constitutes a "floor, not a ceiling" (Stec, & Casey-Lefkowitz, 2000, pp. 45-47). States, at any time, have the right to provide for broader access to information, more extensive public participation in decision-making and a wider access to justice in environmental matters than required by the Convention. The Convention sets forth few requirements that Parties must meet, in order to provide the basis for global and international environmental democracy, namely the effective recognition of aforementioned procedural rights in environmental matters.

---

241 "It is notable that the Aarhus Convention makes no comparable attempt to broaden participation. The real emphasis in the Aarhus Convention is on the involvement of NGOs. However, we should always be aware of the dangers of claiming that NGOs ‚represent' anybody, and of the possibility that a small (even if larger than before) number of participants will wrap up important decisions. More generalized public participation of course faces real obstacles". Lee, & Abbot, 2003, p. 107-108.

FINAL CONCLUSION

"As Popper stated, in The Open Society and its Enemies: '(in our social world) many mistakes would be made which could be eliminated only by a long and laborious process of small adjustments; in other words, by that rational method of piecemeal engineering'. The question is, whether the environment has enough time for such a learning process" (Krämer, 2008, p. 7).

This book started by saying that we are at war, or better, at ecological war between man and nature, between biosphere and "technosphere", humanity against itself.

In this context the purpose was to answer the questions which were presented in the introduction: how can the ecological war be stopped? How can the environmental crisis be resolved?

Or, from a legal viewpoint, the question could be how can political and legal structures contribute to avoid environmental damage and threats of an ecological crisis? How could one begin to reform and restructure actual political institutions so that they are in line with environmental considerations? How can States and their citizens act and organise themselves to provide answers to the current ecological crisis?

The book has suggested a theoretical solution, namely, the transformation of all different forms of governance into an environmental democracy; and it has explored whether the theoretical model already exists, totally or partially, at a global or local level.

In order to reach this objective, it has studied the elements, form, space and actors, on which this new form of democracy is based. To understand form and space it was first necessary to explore the meaning of the terms "democracy" and "environment": what kind of democracy could better achieve the environmental goals, and what kind of definition and approach to the term "environment" could be used to better protect it.

With regard to the form, it is still essential to maintain a democratic model, but it is also indispensable to modify existing representative democracies, not by a radical change as authoritarian or anarchical views suggest, but by the introduction of participatory and deliberative elements.

It has moreover been argued that, although deliberation and participation are distinct and independent elements, the radical democratisation of democracy, which is also crucial for the reduction of the deficit of legitimacy of contemporary politics, can succeed in such instances where participation and deliberation are regarded as the two key elements in the process of collective decision-making.

From an environmental viewpoint, these forms of democracy are likely to be able to achieve environmental goals and build a new form of democracy. The reasons to opt for this form and type of democracy are that, on the one hand, it responds in a better manner to the ecological crisis due to the fact that it introduces environmental values in governmental decisions; on the other hand, it enhances the role of environmental citizens, defining them as controllers of decisions as well as of decision-makers in the environmental field.

Moreover, environmental democracy shall begin a gradual transformation in the form, style and content of democracy; it shall grant environmental rights of citizens; it shall introduce new duties with regard to society's relationship with the rest of nature.

The introduction of environmental rights shall, hence, be balanced with ecological duties: man needs to rediscover respect for Earth because, since he has stopped following the rules of the biosphere and has constructed another world – "technosphere" – he has been cutting himself off from natural rules and he has created new rules which don't match with the planet's rules. So to equilibrate such a dysfunctional world, legal instruments must recognise and implement ecological duties to develop a major individual holistic awareness *vis-à-vis* the Planet.

This has led to the conclusion that a mere traditional citizenship tends to fragment these Earth-oriented responsibilities; it is therefore necessary go beyond the mere recognition of rights and state responsibility for the environment as a concept to involve people in actions concerning global and local environmental problems.

Setting up systems of environmental rights and ecological duties entails the promotion of a new concept of citizenship: environmental and ecological citizenship. For this reason a particular emphasis has been placed on the substantive and procedural rights which possess a link to environmental citizenship, and on those ecological duties which have a more important link to ecological citizenship.

Such a new notion of citizenship derives from the idea that there are specific rights and duties *vis-à-vis* citizens, and in the case of the environment, also rights and duties *vis-à-vis* present and future generations and Planet-Earth.

Concerning the notion of environment, it is possible that as environmental citizens' awareness increases through rights and duties, a less anthropocentric view of environment may be adopted in the future. This could also entail that the decisions start to consider and value any potential impact on the environment, using a more ecocentric approach. The short-term considerations of human welfare will start to be balanced, sometimes even ceded, to long-term interests.

Consequently, the definition of the term "environment", traditionally based on the anthropocentric approach, has to shift to an ecocentric perspective and has to be related to all the various life forms. This way, the goods that have to be protected by environmental law are not the single elements of environment but the results of their interactions.

From a spatial viewpoint, in order to a conduct a radical change in the structure of modern governance, the shift towards environmental democracy shall be made both at the global and local levels.

In the light of the theoretical construction of environmental democracy and its elements, this book examined environmental democracy at the global level by referring to international legal instruments.

In particular, it has explored the elements of the theoretical construction of environmental democracy, examining the notion of "democracy", "environment" and "actors", which can be found in international environmental law, and which have been incorporated into the Aarhus Convention, especially concerning access to information, public participation in decision-making and access to justice in environmental matters. This agreement, viewed from a global perspective, provides an important step towards the construction of environmental democracy.

The Aarhus Convention, thus, was chosen as an empirical effort to move in this direction at a global level, due to the fact that it concretises some fundamental aspects and elements of the theoretical construction of environmental democracy; namely: form, space and actors, although the approach to the environment is still mainly anthropocentric.

Within a formal dimension, the Treaty confirms the role of democracy to achieve environmental goals and provides ways to introduce mechanisms of participatory and deliberative democracy.

On a spatial scale, the Convention focuses on global and local levels. It aims at avoiding the problem of democratic deficit, at democratising not only at the national or sub-national level, but also supra-national and international decision-making processes. In fact, even though the Treaty has a regional geographical scope, it has the potential to spread transparency, openness and democracy into different forms of governance at different levels and thus to enhance the environmental democracy model well beyond the limits of the European region.

Moreover, concerning the actors, the Convention can be viewed as a step towards the creation of a new form of citizenship, which is detached from national boundaries and encompasses rights of the environmental citizen as well as duties of the ecological citizen. It emphasises the role of the citizen, thus, as an individual or in the form of an NGO, providing her/him with substantive and procedural environmental rights and duties.

Concerning environmental rights, the Aarhus Convention has also been identified as a measure of realising links between environmental protection and human rights norms and it represents a contribution to a substantive human right to the environment.

In fact, the focus on procedural rights, as an attempt to facilitate a substantive human right to a healthy environment, offers a pioneering approach, because it demonstrates the increased opening of international mechanisms with regard to the involvement of citizens.

Thus, special attention has been dedicated to the examination of the three pillars, corresponding to the environmental procedural rights to access to information, public participation and access to justice in environmental matters; additionally, an analysis has been provided of the two additional pillars corresponding to the possibility of enforcement of environmental law (fourth pillar), and the review of compliance mechanism (fifth pillar).

The first three pillars are essential to achieve the right to a healthy environment, and to grant the possibility for individuals to be held responsible towards others.

Regarding the fourth pillar, it clarifies that it is not only the environmental authorities and public prosecutors who enforce environmental law, but it also states that the public has a role to play to fulfil their environmental duty to conserve and protect the environment. Concerning the fifth pillar, the compliance mechanism plays a significant role in promoting environmental justice and even in contributing to the implementation of democratic governance.

Linked to this ecological duties perspective, it has been remarked that the Aarhus Convention represents a step forward in the quest of strengthening citizens' environmental rights and duties and with regard to the matching of rights with duties.

In this respect, the rights and duties contained in the Aarhus Convention have the potential to actively improve environmentally responsible, individual decisions and to enhance environmental safeguards.

In conclusion, this book does not claim to have established definite answers and solutions to resolve the ecological crisis and to stop the ecological war but, on the one hand, it suggests how to construct a sustainable legal model to make peace with the Earth; on the other hand, it has explored whether this model already exists at a global and local scale.

The aim is to justify why this model could increase the possibility of meeting environmental goals and why the steps already made, at different levels, are unfortunately still very modest compared to the needs and challenges in the field of environment.

It can be affirmed, hence, that environmental democracy is not a magic formula that solves all environmental problems. However, if the theoretical construction is applied, it can lead to a possible improvement of the assessment of decisions which need to be taken for the protection of Earth, and consequently, for human survival.

Nevertheless, as long as people's preferences are not heavily skewed towards the reduction of all types of environmental degradation, the real positive effect of democracy on many environmental degradation issues will still remain relatively limited.

Thus, to move towards environmental democracy, it is necessary, on the one hand, to improve the already existent political legal structures as well as the participation processes, and, on the other hand, it is also necessary to create a major shift in the awareness of the central significance of the Earth.

Index

Bibliography

1. Books and Articles

Accame, S. (1998). La prima assemblea politica del mondo occidentale". In A. D'Atena, & E. Lanzillotta (Eds.), Alle radici della democrazia. Dalla polis al dibattito costituzionale contemporaneo (p. 11), Roma, Carocci.

Aderson, K. (2005). Book Review: Squaring the Circle? Reconciling Sovereignty and Global Governance through Global Government Networks: A New World Order. Harvard Law Review, 1255.

Agarwal, A., & Narain, S. (1992). A proposal for global environmental democracy. Earth Island Journal, 7, 1.

Allegri, M. R. (2008). Democracy at Union Level: an open question. Political Perspectives, 2, 1

Allen, R. (2000). The New Penguin English Dictionary. London, Penguin.

Alston, P. (1982). A third Generation of Solidarity Rights. Netherlands International Law Review, 307

Alves, C. M. (2003). La protection intégrée de l'environnement en droit communautaire". Revue Juridique de l'Environnement, 129.

Anaya, S. J. (2004). Indigenous Peoples in international law. New York, Oxford University Press.

Anderson, M. R. (1996). Human Rights Approaches to Environmental Protection: An Overview. In: A. Boyle, & M. Anderson (Eds.), Human Rights Approaches to Environmental Protection (p.1)., Oxford, Oxford University Press.

Annan, K. (2000). Foreword. In: S. Stec, S. Casey-Lefkowitz (Eds.), (p. 17). New York, Geneva, United Nations Publication.

Anton, D. K. (1993). The Internationalization of Domestic Law: The Shrinking Domaine Réservé". American Society of International Law Proceedings, 553.

Anton, D. K. (2008). Observations about expanding public participation in the international environmental law-making process. Public Law and Legal Theory Working Paper, Series Working paper n. 112, 8, June 2008. Available at www.ssrn.com/abstract=1145066.

Arblaster, A. (1994). Democracy. Buckingham, Open University Press.

Arditi, S. (2010). Enforcing Individual Producer Responsibility. American Society of International Law Proceedings 2010 Newsletter # 56 European Environmental Bureau, Metamorphosis, also on line available at www.eeb.org/?LinkServID=ADC892EB-C574-5800-28F68DA461162AA2&showMeta=0

Arend, A. C. (1999). Legal Rules and International Society. Oxford, Oxford University Press.

Aristotele, (1946). The Politics of Aristotle. Oxford, Oxford University Press.

Armstrong, K.A. (2008). Civil Society and the White Paper–Bridging or Jumping the Gaps, available at www.jeanmonnetprogram.org

Arrhenius, G. (2007). The boundary problem in democratic theory, Stockholm, available at www.people.su.se/*guarr/texter/boundary. pdf

Attfield, R. (1983). The Ethics of Environmental Concern. Athens, University of Georgia Press.

Attfield, R. (2003). Environmental Ethics, Polity. Cambridge, Wiley.

Baber, W. F., & Bartlett, R. V. (2005). Deliberative Environmental Politics. London, MIT Press.

Baber, W. F., & Bartlett, R. V. (2009). Global Democracy and Sustainable Jurisprudence. London, MIT Press

Bacqué, M.H., Rey, H., & Sintomer, Y. (2005). Gestion de proximité et démocratie participative. Une perspective comparative. Paris, La découverte.

Badni, G. (1993). The Right to Environment in Theory and Practice: The Hungarian Experience. Connecticut Journal of International Law. 439.

Balck, E. C. (2010). Climate Change Adaptation: Local Solution for a Global Problem. Georgetown International Environmental Law Review, 359.

Baldwin, L. D. (1956). Best Hope of Earth, A Grammar of Democracy. Pittsburgh, University of Pittsburgh Press.

Ballesteros, M., & Luk S. (2010). The impact of the Lisbon Treaty–an environmental perspective. available at www.clientearth.org/.

Barber, B. (1984). Strong Democracy: Participatory Politics for a New Age. London, California University Press.

Barker, E. (1942). Reflections on Government. Oxford, Oxford University Press.

Barker, M. J.(1970). The Environmental Citizenship Where to Begin. Art Education, 23, 33.

Barresi, P. A. (1997). Beyond fairness to future generations: An intergenerational alternative to intergenerational equity in the intergenerational environmental arena. Tulaine Environmental Law Journal. 11, 3.

Barry, B. (1978). Circumstances of justice and future generations. In: R. Sikora, & B. Barry (Eds.), Obligations to future generations (p. 204). Philadelphia, White Horse Press.

Barry, H. (2002). Democracy and global warming. London, Biddles Ltd, Guildford and King's Lynn.

Barry, J. (1999). Rethinking Green Politics. London, SAGE.

Barry, J. (2002). Vulnerability and virtue: democracy, dependency, and ecological stewardship. In: B. A. Minteer, & B. Pepperman Taylor (Eds.), Democracy and the Claims of Nature: Critical Perspectives for a New Century (p. 133). Oxford, Rowman & Littlefield Publishers Inc.

Barry, J. (2006). Resistance is fertile: from Environmental to Sustainability Citizenship. In: A. Dobson A., D. Bell (Eds.), Environmental Citizens (p. 21). Cambridge, MIT Press.

Barry, J., Baxter, B. & Dunphy, R. (2004). Europe, Globalisation and Sustainable Development. London, Routledge.

Barstow, Magraw, D., Hawke, L. D. (2007). Sustainable Development. In: D. Bodansky, J. Brunné (Eds.). International Environmental Law (p. 614). Oxford, Oxford University Press.

Bartenstein, K. (2005). Les origines du concept de développement durable. Revue Juridique de l'Environnement, 3, 294.

Barton, B. (2002). Underlying Concepts and Theoretical Issues in Public Participation in resource Development. In: D. Zillman (Ed.), Human Rights in Natural Resource Development: Public participation in the Sustainable Development of Ming and Energy Resources (p. 84). Oxford, Oxford University Press.

Bates, D. C. (2002). Environmental refugees? Classifying human migrations caused by environmental change. Population and Environment, 465.

Beck, U. (1998). Politik der Globalisierung. Frankfurt, Suhrkamp.

Becker, M. L. (1993). The International Joint Commission and Public Participation: Past Experiences, Present Challenges, Future Tasks. Natural Resources Journal, 235.

Beckerman, W. (1999). Sustainable Development and Our Obligations to Future Generations. In: A. Dobson (Ed.), Fairness And Futurity (p. 85). Oxford, Oxford University Press.

Beckman, L. (1994). Democracy and future generations. Why tomorrow's people should not vote today. In: M. Coppens, A. Gosseries, & J-C. Merle (Eds.), Intergenerational justice, Oxford, Oxford University Press.

Beckman, L. (2006). Citizenship and voting rights: should resident aliens vote?. Citizenship Studies, 10 (2), 153.

Beckman, L. (2007). Democracy, future generations and global climate change. In: Prepared for the workshop "Democracy on the day after tomorrow" at the ECPR Joint Sessions, Helsinki.

Beigbeder, Y. (1992). Le rôle international des organisations non gouvernementales. Paris, Bruylant.

Bell, D. R. (2004 a). Environmental Refugees: What Rights? Which Duties. Res Publica, 135.

Bell, D. R. (2004 b). Sustainability through democratisation? the Aarhus convention and the future of environmental decision-making in Europe. In: J. Barry, B. Baxter, & R. Dunphy (Eds.), Europe, Globalisation and the Challenge of Sustainability (p. 94). London, Routledge.

Bell, D. R. (2005). Liberal Environmental Citizenship. Environmental Politics, 14(2), 179.

Belrhali-Bernard, H. (2009). Le droit de l'environnement: entre incitation et contrainte. Revue Droit Public, 1683.

Benson, D., & Jordan, A. (2008). A Grand Bargain or an Incomplete Contract? European Union Environmental Policy after the Lisbon Treaty. European Energy and environmental Law Review, 280.

Benvenisti, E. (2005). The Interplay between Actors as a Determinant of the Evolution of Administrative Law in International Institutions. Law and Contemporary Problems, 319.

Berge, E. (1994). Democracy and Human Rights: Conditions for Sustainable Resource Utilisation. In: B.R. Johnson (Ed.), Who Pays the Price? The Socio cultural Context of Environmental Crisis (p. 187), Covelo. Island Press.

Bergesen, H.O., & Parmann, G. (1992). In Green Globe Yearbook of International Cooperation on Environment and Development. Oxford, Oxford University Press.

Bergkamp, L. (2001). Liability and Environment – Private and Public Law Aspects of Civil Liability for Environmental Harm in an International Context. The Hague, Kluwer Law International.

Bergkamp, L. (2002). Corporate Governance and Social Responsibility: a New Sustainability Paradigm?. European Environmental Law Review, p. 136.

Bernstein, S. (2005). Legitimacy in Global Environmental Governance. Journal of International Law and International Relations, 139.

Bessette, J. (1980). Deliberative Democracy: The Majority Principle in Republican Government. In: How Democratic is the Constitution? (p. 102). Washington, American Enterprise Institute.

Besson, S., & Utzinger A. (2007). Introduction: Future Challenges of European Citizenship: Facing a Wide-Open Pandora's Box. European Law Journal, 573.

Betlem, G., Brans E.H.P. (2002). The Future Role of Civil liability for Environmental Damage in the EU. Yearbook of European Environmental Law, 2, 183.

Birch A. H (1993). The Concepts and Theories of Modern Democracy. London, Routledge.

Birkin-Shaw, P. (2004). A Constitution for the European Union? A Letter from Home. European Public Law, 57.

Birkin-Shaw, P. (2006). Freedom of Information and Openness: Fundamental Human Rights. Administration Law Review, p. 177.

Birnie, P., Boyle A. (2002). International Law and the Environment. Oxford, Oxford University Press.

Birnie, P., Boyle A., Redgwell, C. (2009). International law and the environment. Oxford, Oxford University Press.

Bjerler, N. (2009). Do Europeans Have a Right to Environment?", available at www.esil-sedi.eu/fichiers/en/Bjerler_455.pdf, 2009.

Blanc-Jouvan, X. (1971). Problems of harmonisation of traditional and modern concepts in the land law of french-speaking Africa and Madagascar, integration of customary and modern legal systems in Africa. New York, African Pub. Corp.

Bobbio, N. (1978). Democrazia rappresentativa e democrazia diretta. In: G. Quazza (Ed.), Democrazia e partecipazione (p. 22). Torino, Giappichelli.

Bobbio, N. (1984). Il futuro della democrazia. Una difesa delle regole del gioco. Torino, Giappichelli.

Bobbio, N. (1997). L'età dei diritti. Torino, Einaudi.

Bodansky, D. (2007). Legitimacy. In: Bodansky D., & J. Brunnée (Eds.), International Environmental Law (p. 704). Oxford, Oxford University Press.

Bodansky, D. (2009). Is There an International Environmental Constitution?. Indiana Journal of Global Legal Studies, 16, 565.

Bohman, J. (1996). Public Deliberation: Pluralism, Complexity and Democracy. Cambridge, MIT Press.

Boiteux, M. (2003). L'homme et sa planèt. Paris, Académie des sciences morales et politiques.

Bolton, J. R. (2000). Should We Take Global Governance Seriously?. Chicago Journal of International Law, 205.

Bonine, J. E. (2003). The public's right to enforce environmental law. In: S. Stec (Ed.), Handbook on Access to Justice under the Aarhus Convention (p. 31), Szentendre, Unites Nation Publication.

Bookchin, M. (1988). Toward and Ecological Society. California, Black Rose Books

Bookchin, M. (1990). Ecology and Revolutionary Thought. In: M. Bookchin (Ed.), Post-Scarcity Anarchism. San Francisco, AK Press.

Boon, E. K., & Le Tran, T. (2007). Are Environmental Refugees Refused?. Studies of Tribes and Tribals, 89.

Borloo, J.L. (entretien avec) (2008). Le développement durable n'est plus une question parmi d'autres mais bien une préoccupation placée au cœur de toutes les autres. Les Petites Affiches, 22 avr. 81, 9.

Bosselmann, K. (2008). The Principle of Sustainability. Aldershot, Ashgate Publishing Company.

Bosselmann, K. (2009). The Way Forward: Governance for Ecological Integrity. In: L. Westra, K. Bosselmann, & R. Westra, (Eds.), Reconciling Human Existence with Ecological Integrity (p. 319). London, Earthscan.

Boyle, A. E. (1996). The role of International Human Rights Law in the Protection of the Environment. In: A. Boyle, & M. Anderson (Eds.), Human rights Approaches to environmental protection (p. 43). Oxford, Oxford, Oxford University Press.

Bradford, M. (1996). Protecting the environment for future generations: A proposal for a republican superagency. New York University Environmental Law Journal, 5.

Braibant, G. (2003). L'environnement dans la Charte des droits fondamentaux de l'Union européenne. Les Cahiers du Conseil Constitutionnel, 15, 262.

Brandl, E., & Bungert, H. (1992). Constitutional Entrenchment of Environmental Protection: A Comparative Analysis of Experiences Abroad. Harvard Environmental Law Review, 16, 4.

Breitmeier, H., & Rittberger, V. (2000). Environmental NGOs in an Emerging Global Civil Society. In: P. S. Chasek (Ed.), The Global Environment In The Twenty-First Century: Prospects For International Cooperation (p. 130). New York, Institut für Politikwissenschaft.

Breton-Le Goff, G. (2001). L'influence des organisations non gouvernementales (ONG) sur la négociation de quelques instruments internationaux. Brussels, Bruylant.

Brooke, D. (2006). Hall Memorial Lecture, Environmental Justice: The Cost Barrier. Journal of Environmental Law, 341.

Bruch, C.E., & Czebiniak, R. (2002). Globalising Environmental Governance: Making the Leap from Regional Initiatives on Transparency, Participation, and Accountability in Environmental Matters, XXXII. Environmental Law Reporter, 1428.

Bugge, H. C., & Voigt, C. (2008). Sustainable Development in International and National Law. Groningen, Europa Law Publishing.

Bullard, R. D. (1996). Unequal protection: Environmental justice and communities of color. San Francisco, Sierra Club Books.

Burhenne, W. E., & Tarasofsky, R. G. (1998). Codification and Progressive Development of International Law – An Example from the Field of the Environment. Environmental Policy and Law, 77.

Burker, M., & Rees, A. (1996). Citizenship Today: The Contemporary Relevance of T.H. Marshall. London, University of London Press.

Burnett-Hall, R., & Jones, B. (2009). Environmental Law. London, Thomson Reuters.

Butler, J., & De Schutter, O. (2008). Binding the Eu to International Human Rights Law. Yearbook of European Law, 277.

Caldwell, L. K. (1980). International Environmental Policy and Law. Durham, Duke University Press.

Callicott, J.B. (1980). Animal Liberation: A Triangular Affair. Environmental Ethics, 2, 311.

Cameron, J. (1996). Compliance, Citizens and NGOs. In: J. Cameron, J. Werksman, & P. Roderick (Eds.), Improving Compliance With International Environmental Law (p. 29). London, Earthscan Pub.

Cameron, J., & Mackenzie, R. (1996). Access to Environmental Justice and Procedural Rights in International Institutions., In: A. Boyle, & M. Anderson (Eds.), Human Rights Approaches to Environmental Protection (p. 129). Oxford, Oxford University Press.

Campiglio, L., Pineschi, L., Siniscalco, D. & Trevest, T. (1994). The Environment After Rio. Dordrecht, London, Graham & Trotman/martinus Nijhoff.

Cancado Trindade, A. A. (1992). The contribution of international human rights law to environmental protection, with special reference to global environmental change. In: E. B. Weiss (ed.), Environmental Change and International Law (p. 244). Hong Kong, United Nations University Press.

Caney, S., & Simons, C. (2005). Cosmopolitan justice, responsibility, and global climate change. Leiden Journal of International Law, 18, 747.

Cano, G. J. (1975). A Legal and Institutional Framework for Natural Resources Management. FAO Legislative Studies, Rome, 9, 1.

Cano, G. J. (1975). A Legal and Institutional Framework for Natural Resources Management. Rome.

Cantat, O. (2008). Développement durable: une pensée de référence difficile à mettre en œuvre. Droit de l'Environnement, 1 juillet , n° spéc.

Caranta, R. (1993). Governmental liability after Francovich. Cambridge Law Journal, 272.

Caranta, R. (2008). Interest representation in administrative proceeding. Napoli, Jovene.

Carolan, M. (2006). Ecological representation in deliberation: the contribution of tactile spaces. Environmental Politics,15 (3), 345.

Carson, R. (1962). Silent Spring. Boston, Houghton Mifflin Harcourt ,

Cassin, R. (1974). Les droits de l'Homme, IV Recueil des Cours, 323, 327. Leyden.

Ceiner, G. (1984). Porte, Portici e Logge. Udine.

Chambers, N., Simons, C., & Wackeragel, M. (2000). Sharing Nature's Interest: Ecological Footprints as an Indicator of Sustainability. London, Earthscan.

Chambers, S. (2003). Deliberative democratic theory., Annual Review of Political Science, 6, 307.

Chamboredon, A. (2007). Du Droit de l'Environnement au droit à l'Environnement, A la recherche d'un juste milieu. Paris, L'Harmattan.

Charney, J. I., Anton, D. K. & O'Connell, M. E. (Eds.), Politics, Values And Functions: International Law In The 21st Century: Essays In Honor Of Professor Louis. Henkin. Martinus Nijhoff Publishers.

Charnovitz, S. (1997). Two Centuries of Participation: NGOs and International Governance. Michigan Journal of International Law, 183.

Charnovitz, S. (2003). The Emergence of Democratic Participation in Global Governance (Paris 1919). Indiana Journal of Global Legal Studies, 45.

Charnovitz, S. (2006a). Centennial Essay: In Honour of the Tenth Anniversary of the AJIL and the ASIL. American Journal of International Law, 348.

Charnovitz, S. (2006b). Nongovernmental organisations and International Law. American Journal of International Law, 348.

Chekki, D. A. (1979). Participatory Democracy in Action: International Profiles of Community Development. Bombay, Vikas Publishing House.

Chioma Steady, F. (2009). Environmental justice in the new millennium: global perspectives on race, ethnicity and human rights. New York, Palgrave Macmillan.

Chiti, E. (2002). Legittimazione ad agire ex art. 230 del Trattato ed effettività della tutela giurisdizionale. Giornale di diritto amministrativo, 11, 1169.

Chiti, E. (2005). The Relationship between National Administrative Law and European Administrative Law in Administrative Procedures. In: J. Ziller (Ed.), What's New in European Administrative Law?, European University Institute, Department of Law, Working Paper , 10, 7.

Chiti, E. (2010). Trattato di Lisbona. La Cooperazione Amministrativa. Giornale di diritto amministrativo, 241.

Christoff, P. (1996). Ecological citizens and ecologically guided democracy. In: B. Doherty, & M. De Geus (Eds.), Democracy and Green Political Thought. Sustainability, Rights and Citizenship (p. 151). London, Routledge.

Clarke, P.B. (1999). Deep Citizenship. London, Pluto Press.

Coenen, F. (2008), Public Participation and Better environmental Decisions. Enschede, Springer

Coffey, C., & Newcombe, J. (2001). The Polluter Pays Principle and Fisheries: the Role of Taxes and Charges. London, Routledge.

Cohen, J. (1989). Deliberative Democracy and Democratic Legitimacy. In: A. Hamlin, & P. Pettit, (Eds.), The Good Polity (p. 17). Oxford, John Wiley & Sons, Limited.

Cohen, J., & Rogers, J. (2003). Power and Reason. In: A. Fung, & E. Olin Wright (Eds.), Deepening Democracy: Institutional Innovations in Empowered Participatory Governance (p. 237). New York, Princeton University Press.

Cohen, M. & Murphi, J. (2001). Exploring sustainable consumption: environmental policy and the social sciences. Oxford, Emerald Group Publishing Limited.

Collins, L. (2006). The Constitution: A Carter for Sustainable development in Europe?. In: M. Pallemaerts, & A. Azmanova (Eds.), The European Union and Sustainable Development: Internal and External Dimension (p. 93). Brussels, ASP-VUB Press.

Collins, L. (2007a). Are We There Yet? The Right to Environment International and European Law. McGill International Journal of Sustainable Development Law and Policy, 120.

Collins, L. (2007b). Environmental Rights for the Future? Intergenerational Equity in the EU. Review of European Community and International Environmental Law, 16, 321.

Comba, D. (2009). Prochaine entrée en vigueur du protocole de Kiev sur les registres des rejets et transferts de polluants. Sentinelle 27 September, Available at www.sfdi.org/actualites/a2009/Sentinelle%20197.htm#kiev

Commoner, B. (1992). Making peace with the planet. New York, The New Press.

Comte, F. (2003). Criminal Environmental Law and Community Competence. European Environmental Law Review, 147.

Congress on Public International Law, (1995) Environmental Policy and Law, p. 163.

Cooper, D. E., & Palmer, J. (1998). Spirit of the Environment. London, Routledge.

Corazza, C. (2009). EcoEuropa, Le nuove politiche per l'energia e il clima. Milano, EGEA.

Cordonier Segger, M.-C & Weeramantry, C.G. (2005). Sustainable Justice: Reconciling Economic, Social and Environmental Law. London, Martinus Nijhoff.

Cordonier Segger, M.C. (2004). Significant Developments in Sustainable Developments Law and Governance: A Proposal. Natural Resources Forum, 28,61.

Craig, P. (1991). Francovich, remedies and the scope for damages liability. Law Quarterly Review, 595.

Craig, P. (1997). Democracy and Rule-making Within the EC: an empirical and normative assessment. European Law Journal, 105.

Craig, P. (1999). The Nature of the Community: Integration, Democracy and Legitimacy. In: P. Craig, & G. De Burca (Eds.), The Evolution of EU Law (p. 41). Oxford, Oxford University Press.

Craig, P., & De Burca, G. (2008). EU Law. Text, cases and materials. Oxford, Oxford University Press.

Craig, P., & De Burca, G. (1999). The Evolution of EU Law. Oxford, Oxford University Press

Cramer, B. W. (2009). The Human Right to information, the environment and information about the environment: from the Universal Declaration to the Aarhus Convention. Communication Law and Policy, 14, 73.

Cremona, M. (2003). The Draft Constitutional Treaty: external relations and external action. Common Market Law Review, 1347.

Cremona, M. (2004). The Union as a Global Actor: roles, models and identity. Common Market Law Review, 553.

Crosetti, A. & Fracchia, F. (2002). Procedimento amministrativo e partecipazione. Problemi, prospettive ed esperienze. Milano, Giuffré.

Cross, G. (1995). Subsidiarity and the Environment. Yearbook of European Law, 107.

Crossent, T., & Niessen, V. (2007). NGO Standing in the European Court of Justice – Does the Aarhus Regulation Open the Door?. Review of European Community & International Environmental Law, 3, 332.

Cullet, P. (1995). Definition of an Environmental Right in a Human Rights Context. Netherlands Q. Human Rights, 25.

D'Amato, A. (1990). Do We owe a Duty to Future generations to Preserve the global Environment?. American Journal of International Law, 190.

Dahl, R. (1956). A Preface to democratic Theory. Chicago, University Of Chicago Press.

Dahl, R. (1991). Modern Political Analysis. London, Pearson.

Dahl, R. (1998). On democracy. Yale, Yale University Press.

Daly, H. E. (1973).Toward a Steady State Economy. San Francisco, W.H.Freeman & Co Ltd.

Dannenmaier, E. (1997). Democracy in Development: Toward a Legal Framework for the Americas. Tulane Environmental Law Journal, 111.

Dannenmaier, E. (2007). A European Commitment to Environmental Citizenship: Article 3.7 of the Aarhus Convention and Public Participation in International Forums. Yearbook of International Environmental Law, 33.

Davies, S. (2007). In Name or Nature? Implementing International Environmental Procedural Rights in the Post-Aarhus Environment: A Finnish Example. Environmental Law Review, 190.

Davis, J. (2007). Conceptual Change", Emerging Perspectives on Learning, Teaching and Technology. University of Georgia. 3 October, 2007. Available at www.projects.coe.uga.edu/epltt/index.php?title=Conceptual_Change.

De Abreu Ferreira, S. (2007). Fundamental Environmental Rights in EU Law. 6th Global Conference (2007) Monday 2nd July – Thursday 5th July 2007, Mansfield College, Oxford, Available at www.inter-disciplinary.net/ptb/ejgc/ejgc6/ Ferreira%20paper.pdf, p.5.

De Abreu Ferreira, S. (2007). The Fundamental Right of Access to Environmental Information in the EC: A Critical Analysis of WWF-EPO v. Council. Journal of Environmental Law, 19, 399.

De Burca, G. (1996). The Quest for Legitimacy in the EU. Modern Law Review, 349.

De Burca, G. (2006). After the Referenda. European Law Journal, 6.

De Geus, M. (1996). The ecological restructuring of the state. In: B. Doherty, & M. De Geus (Eds.) Democracy and Green Political Thought (p. 190). London, Routledge.

De La Fayette, L. (2002). The Concept of Environmental Damage in International Liability Regimes. In: M. Bowman, & A. Boyle (Eds.), Environmental Damage in International and Comparative Law – Problems of Definition and Valuation (p. 149), New York, OUP Oxford.

De Lange, F. (2003). Beyond Greenpeace, Courtesy of the Aarhus Convention, in EC Environmental Law. Yearbook European Environmental law, 227.

De Sadeleer, N. (1999). Les principes du pollueur-payeur, de prévention et de précaution. Bruxelles, Emile Bruylant.

De Sadeleer, N. (2005). Access to justice in environmental matters and the role of NGOs. Groningen, Europa Law Pub.

De Schutter, O. (1996). Sur l'émergence de la société civile en droit international: la Cour européenne des droits de l'homme. European Journal of International Law, 372.

Dean, H. (2001). Green Citizenship. Social Policy and Administration, 35, 490.

Déjeant-Pons, M. (1999). La Convention de Berne relative à la conservation de la vie sauvage et du milieu naturel en Europe. In: Déjeant-Pons, M. (Ed.): Vers l'application renforcée du droit international de l'environnement/Towards strengthening application of international environmental law (p. 58). Paris, Edition Frison Roche.

Déjeant-Pons, M. (2002). Human Rights to Environmental Procedural Rights. In: M. Déjeant-Pons, & Pallemaerts M. (Eds.), Human Rights and the Environment (p. 23). Strasbourg, Council of Europe.

Del Rey, M.-J. (2010). « Développement durable »: l'incontournable hérésie. Droit, 1493.

Delreux, T. (2009). The Eu in Environmental Negotiations in UNECE: An Analysis of its Role in the Aarhus Convention and the SEA Protocol Negotiations. Review of European Community and International Environmental Law, 328.

Descartes, R. (1931). Discourse on Method. In: The philosophical Works of Descartes. Cambridge, Cambridge University Press.

Desgagne, R. (1995). Integrating Environmental Value into the European Convention on Human Rights. Am Journal international Law, 263.

Dette, B. (2004). Access to Justice in Environmental Matters. In: M. Onida (Ed.), Europe and the Environment – Essays in Honour of Ludwig Krämer (p. 3), Groningen, Europa Law Publishing.

Devall, B., & Sessions, G. (1984). The Development of Natural Resources and the Integrity of Nature. Environmental Ethics, 6, 296.

Dobson, A. (1995). Green political thought. London, Routledge.

Dobson, A. (1996). Representative democracy and the environment. In: W.M. Lafferty, & Meadowcroft, J. (Eds.), Democracy and the environment: problems and prospects (p. 125), Cheltenham, Edward Elgar Publishing Ltd.

Dobson, A. (1998). Justice and the Environment: Conceptions of Environmental Sustainability and Dimensions of Social Justice. Oxford, Clarendon Press.

Dobson, A. (2000). Ecological citizenship: a disruptive influence?". Available at www.vedegylet.hu/okopolitika/Dobson%20-%20Ecological%20Citizenship.pdf

Dobson, A. (2003). Citizenship and the Environment. London, OUP Oxford.

Dobson, A. (2004). Social inclusion, environmental sustainability and citizenship education. In: J. Barry, B. Baxter, & R. Dunphy (Eds.), Europe, Globalisation and Sustainable Development (p. 115), London, Routledge.

Dobson, A. (2005). Citizenship. In: Dobson, A., & Eckersley, R. (Eds.), Political Theory and the Ecological Challenge (p. 481). Cambridge, Cambridge University Press.

Dobson, A. 1999), Fairness And Futurity. Oxford, OUP.

Dobson, A., & Bell, D. (2006). Environmental Citizens. Oxford, OUP.

Dobson, A., & Bell, D. (2006). Introduction. In: A. Dobson., & D. Bell (Eds.), Environmental Citizenship (p. 1). Cambridge, Cambridge University Press.

Dobson, A., & Saiz, A.V. (2005). Introduction. Environmental Politics, 157.

Dodeller, S., & Pallemaerts, M. (2005). L'accès des particuliers à la Cour de Justice et au Tribunal de Première Instance des Communautés européennes en matière d'environnement: bilan du droit positif et perspectives d'évolution. In: C. Larssen, & M. Pallemaerts (Eds.), L'accès à la justice en matière d'environnement/Toegang tot de rechter in milieuzaken (p. 287). Brussels, Emile Bruylant.

Donald, K. A. (2008). Observations About Expanding Public Participation In The International Environmental Law-Making Process. The Social Science Research Network Electronic Paper Collection. Available at www.ssrn.com/abstract=1145066.

Donnelly, B., & Bishop, P. (2007). Natural Law and Ecocentrism. Journal of Environmental Law, 89.

Douglas-Scott, S. (1996). Environmental Rights In the European Union – Participatory Democracy or Democratic Deficit?. In: Boyle, A., & Anderson, M. (Eds.), Human rights Approaches to environmental protection (p. 109). Oxford, Oxford, Oxford University Press..

Dozer, R. (1976). Property and Environment: The Social Obligation Inherent in Ownership. Morges, International Union for Conservation of Nature.

Drengson, A.R. (1998). Shifting Paradigms: from the technocratic to the person-planetary. Environmental Ethics, 221.

Drevensek, M. (2005). Negotiation as the Driving Force of Environmental Citizenship. Environmental Politics. 14(2), 226.

Dryzek, J. (2000). Deliberative democracy and beyond: liberals, critics, contestations. Oxford.

Dunoff, J. L. (2007). Levels of Environmental Governance. In: D., Bodansky, & J. Brunnée (Eds.), International Environmental Law (p. 85). Oxford, Oxford University Press.

Dwivedi, O. P. (2006). Hindu Religion and Environmental Well-Being. In: R. S. Gottlieb (Ed.) Religion and Ecology (p. 160). Oxford, Routledge.

Ebbesson, J. (2006). Access to justice at the level of the EU: progress or stagnation?. In: A. Gourtin, (Ed.) The Aarhus Regulation: New Opportunities for Citizens in the EU and Beyond?. Report of the ECOSPHERE Forum held in Brussels, 27 October 2006, Brussels, ECOSPHERE.

Ebbesson, J. (2009). Environmental law and justice in Context. Cambridge, Cambridge University Press.

Ebbesson, J.(2007). Public Participation. In: D. Bodansky,& J. Brunnée (Eds.), International Environmental Law (p. 683). Oxford, Oxford University Press.

Eberhard, C. (2008). Traduire nos responsabilités planétaires, recomposer nos paysages juridiques. Bruxelles, Emile Bruylant.

Eckersley, R. (1992). Environmentalism and Political Theory. London, Routledge.

Eckersley, R. (1996). Greening Liberal Democracy: The Rights Discourse Revisited. In: B. Doherty, & M. De Geus (Eds.), Democracy and Green Political Thought (p. 214), London, Routledge.

Eckersley, R. (2000). Deliberative Democracy, Ecological Representation and Risk: Towards a Democracy of the Affected. In: M. Saward (Ed.), Democratic Innovation: Deliberation, Representation and Association (p. 230). London, Routledge.

Eckersley, R. (2001). Ecofeminism and Environmental Democracy: Exploring the Connections. Women & Environments International Magazine, 52.

Eckersley, R. (2004). The Green State: Rethinking Democracy and Sovereignty. Cambridge, MIT Press.

Editorial (2007a). What should replace the Constitutional Treaty?. Common Market Law Review, 561.

Editorial (2007b). Democracy and the Union: Dressing up Cinderella. European Constitutional Law Review, 3, 353.

Egger, R. (2007). Press Officer Atomstopp – Initiative "1 Million against Nuclear Power", June 28.

El-Hinnawi, E. (1985). Environmental Refugees. Nairobi, Kenya: United Nations Environmen- tal Programme..

Elgar, E., Doeleman J., & Sandler, T. (1998). The intergenerational case of missing markets and missing voters. Land Economics, 1.

Ellickson, R. (1991). Order Without Law: How Neighbors Settle Disputes. Harvard, Harvard University Press.

Elliot, R. (1995). Faking Nature. Inquiry. In R. Elliot (Ed.), Environmental Ethics (p. 76), Oxford, OUP Oxford.

Emiliou, N. (1992). Subsidiarity: An Effective Barrier Against the "The Enterprises of Ambition"?. European Environmental Law Review, 383.

Emiliou, N. (1996). State liability under Community law: shedding more light on the Francovich principle?. European Environmental Law Review, 399.

Engin, F., & Turner, S. (2002). Handbook of Citizenship Studies. London, SAGE Publications Ltd.

Epstein, R. (1998). Justice across generations. Texas Law Review, 67.

Estlund, D. (2003). The democracy/contractualism analogy. Philosophy and Public Affairs, 31.

Estyd, C. & Ivanova, M.H. (2002). Global Environmental Governance, Options and Opportunities. New Haven, Yale School of Forestry.

Europa. The E.U. at a glance, Treaties & Law. Available at www.europa.eu/abc/treaties/index_en.htm

Exell Pirro, D. (2008). Introduction Women and an International Court of the Environment. In: A. Postiglione (Ed.), The Protection and Sustainable Development of the Mediterranean Black Sea Ecosystem (p. 833). Bruxelles, Bruylant.

Farrelly, C. (2004). Introduction to Contemporary Political Theory. London, Sage Publications Ltd.

Feinberg, J. (1998). Harmless wrongdoing: The moral Limits of the Criminal Law. Oxford, Oxford University Press.

Fergusson, R., Manser, M., & Pickering, D. (2000). The New Penguin Thesaurus. Harmondsworth, Penguin Books Ltd.

Fievet, G. (2001). Réflexions sur le concept de développement durable: prétention économique, principes stratégiques et protection des droits fondamentaux. Revue belge de droit international, 128.

Finger, M. (2008). Which governance for sustainable development?. In: J. Park, K. Conca, & M. Finger (Eds.), The crisis of Global Environmental Governance (p. 35). London, Routledge.

Finley, M.I. (1973). Democracy Ancient and Modern. New York, Rutgers University Press.

Finley, M.I. (1983). Politics in the Ancient World. Cambridge, Cambridge University Press.

Fish, S. (1999). Mutual respect as a device of exclusion. In: S. Macedo (Ed.), Deliberative politics: essays on democracy and disagreement (88). New York, OUP USA.

Fitzmaurice, M. (2003). Public Participation in the North American Agreement on Environmental Cooperation. International Law & Comparative Law Quarterly, 333.

Fitzmaurice, M. (2009). Contemporary Issues in International Environmental Law. Cheltenham, Edward Elgar Publishing Ltd.

Fitzmaurice, M. (2009). Environmental justice through international complaint procedure? Comparing the Aarhus Convention and the North American Agreement on Environmental Cooperation. In: J. Ebbesson, (ed.), Environmental law and justice in Context (p. 211). Cambridge, Cambridge University Press.

Flynn, B. (2008). Planning Cells and Citizen Juries in Environmental Policy: Deliberation and Its limits. In: F. Coenen (Ed.), Public Participation and Better environmental Decisions (p. 57). Enschede. Springer.

Flynn, R., Bellaby, P., & Ricci, M. (2008). Environmental citizenship and public attitudes to hydrogen energy technologies. Environmental Politics, 17, 766.

Ford K. (1998). Can a Democracy Bind Itself in Perpetuity? Paine, the Bank Crisis, and the Concept of Economic Freedom. Proceedings of the American Philosophical Society, 142.

Foster, J. (1997). Valuing nature? Economics, ethics and environment. London, Routledge

Fowles, B. (2002). Meeting Human and Ecological Rights in Creating the Sustainable Built Environment. 1er Global Conference "Environmental Justice and Global citizenship", 14th – 16th February 2002 Copenhagen. Denmark.

Fox, W. (1989). The Deep Ecology-Ecofeminism Debate and Its Parallels. Environmental Ethics, 5.

Fraccia, F. (2009). The Legal Definition of Environment: From Rights to Duties. Research Paper, n. 2009. Available at www.ssrn.com/abstract=850488

Fralin, R. (1978). Rousseau and Representation. A Study of the Development of his concept of Political Institutions. New York, Columbia University Press.

Francioni, F. (2008). Access to Justice in International environmental law. In: A. Postiglione (Ed.), The Protection and Sustainable Development of the Mediterranean Black Sea Ecosystem (p. 25). Bruxelles, Bruylant.

Francioni, F. (2009). Access to Justice as a Human Right. Oxford, Oxford University Press.

French, H. (1996). The Role of Non-State Actors. In: J. Werksman (Ed.), Greening International Institutions (p. 251). London, Earthscan.

Frumer, P. (1998). Protection de l'environnement et droits procéduraux de l'homme: des relations tumultueuses?. Revue Trimestrielle Droit Homme, 813.

Gamble, A. & Wright, T. (1999), The New Social Democracy. Oxford, Wiley.

Gardner, G., & Stern, P. (2002). Environmental Problems and Human Behaviour. Boston, Pearson Custom Publishing.

Gardner, J. (1978). Discrimination against future generations: The possibility of constitutional limitations. Environmental Law, 9.

Gates, S., Gleditsch, N. P., & Neumayer, E. (2003). Environmental Commitment, Democracy and Inequality, Background Paper, World Development Report 2003, World Bank. Washington. Work Bank Publication.

Gbikpi, B. , & Grote, J. R. (2002). From Democratic Government to Participatory Governance. In: H. Heinelt, & A. Opladen (Eds.), Participatory Governance in Multi-Level Context: Concept and Experience (p. 17). Leverkusen, Leske + Budrich.

Geisinger, A. (2002). A Belief Change Theory of Expressive Law. Iowa Law Review, 88, 35.

Geisinger, A. (2009). Expressive Environmental Regulation: How Law Influences Beliefs About How to Live Sustainably. 8th Global Conference "Environmental Justice and Global Citizenship", 10th – 12th July 2009. Oxford, Oxford University Press.

Gemmill, B., & Bamiele-Izu, A. (2002). The Role of NGOs and Civil Society in Global Environmental Governance. In: C. Estyd, & M.H. Ivanova (Eds.), Global Environmental Governance, Options and Opportunities (p. 77). New Haven, Yale School of Forestry & Environmental Studies.

Gendron, C., Vaillancourt, J.G., & Audet, R. (2010). Développement durable et responsabilité sociale, de la mobilisation à l'institutionnalisation. Quebec, Presses Internationales Polytecnique.

Giagnocavo, C., & Goldstein, H. (1990). Law Reform or World Reform: The Problem of Environmental Rights. McGill Law Journal, 345.

Gillespie, A. (1997). International environmental law, policy, and ethics. Oxford, OUP Oxford.

Gilpin, A. (2000). Dictionary of Environmental Law. Cheltenham, Edward Elgar Publishing Ltd.

Gleditsch, N. P. (1998). Armed Conflict and the Environment: A Critique of the Literature. Journal of Peace Research, 381.

Gleditsch, N. P., & Sverdlop, B.O. (2003). Democracy and the Environment. In: E. Paper, & M. Redclift (Eds.), Human Security and the Environment: International Comparisons (p. 70). London, Edward Elgar Publishing Ltd.

Goodin, R. E. (1992). Green Political Thought. Cambridge, Routledge.

Goodin, R. E. (2003). Reflective Democracy. Oxford, OUP Oxford.

Goodin, R. E. (2007). Enfranchising the all-affected and its alternatives. Philosophy and Public Affairs. 35, 40.

Gormley, W. P. (1976). Human Rights and Environment: The Need For International Co-operation. Amsterdam, Sijthoff.

Gormley, W. P. (1990). The legal Obligation of the International Community to Guarantee a Pure and Decent Environment; The Expansion of Human Rights Norms. Geo International Environmental Law Review, 85.

Gouguet, J.J. (2007). Développement durable et décroissance, deux paradigmes incommensurables. In: Pour un droit commun de l'environnement, Mélanges en l'Honneur de Michel Prieur (p. 124). Paris, Dalloz.

Gourtin, A. (2006). The Aarhus Regulation: New Opportunities for Citizens in the EU and Beyond? Report of the ECOSPHERE Forum held in Brussels, 27 October 2006. Brussels, ECOSPHERE.

Graham, S. (2003). Deliberative Democracy and the Environment. London, Routledge.

Green, J.F. (2004). Engaging the Disenfranchised: Developing Countries and Civil Society in International Governance for Sustainable Development–An Agenda for Research, UNU-IAS Report, United Nations University, Tokyo, UNU/IAS.

Greven, M. TH. (2007). Some Considerations on Participation in Participatory Governance. In: B. Kohler-Kock, & Rittberger, B. (Eds.). Debating the Democratic Legitimacy of the European Union (p. 233), Plymouth, Rowman & Littlefield.

Gros, M. (2009). Quel degré de normativité pour les principes environnementaux?. Revue Droit Public, 425.

Grossman, M.R. (2006). Agriculture and the Polluter Pays Principle: An Introduction. Oklahoma Law Review, 59, 1.

Grubb, M. (1993). The Earth Summit Agreements: A Guide and Assessment. London, Earthscan Publications Ltd.

Guha, R. (1989). Radical American Environmentalism and Wilderness Preservation. A third World Critique. Environmental Ethics, 11, 71.

Gupte, M., & Bartlett, R.V. (2007). Necessary Preconditions for Deliberative Environmental Democracy? Challenging the Modernity Bias of Current Theory. Global Environmental Politics, 94.

Gutmann, A., & Thompson, D. (1996). Democracy and disagreement. Cambridge, Belknap Press of Harvard University Press.

Habermas, J. (1973) Legitimation Crisis. Boston, Beacon Press.

Habermas, J. (1987a). The Philosophical Discourse of Modernity. Cambridge, The MIT Press.

Habermas, J. (1987b). Theory of Communicative Action. Boston, Beacon Press.

Habermas, J. (1991). Citizenship and National Identity: Some Reflections on the Future of Europe. Praxis International,12, 1.

Habermas, J. (1996). Between Facts and Norms. Cambridge, The MIT Press.

Habermas, J. (1998). The inclusion of the Other: Studies in Political Theory. Cambridge, The MIT Press.

Hailwood, S. (2005). Environmental citizenship as reasonable citizenship. Environmental Politics, 14, 195.

Hall, N. D. (2007). The evolving role of citizens in united states-canadian international environmental law compliance. Pace Environmental Law Review, 131.

Hall, P. (1997). Sustainable Cities for Europe. In: V. P. Mega, & R. Petrella (Eds.), Utopias and Realities of Urban Sustainable Development, New Alliances between Economy, Environment, and Democracy for small and Medium-sized Cities (p. 24). Dublin, European Foundation for the Improvement of Living and Working Conditions.

Hallo, R. E. (2007). How far has the EU applied the Aarhus Convention?. Brussels, European Environmental Bureau.

Hallo, R. E. (2008). Unwelcome Guests: Etiquette in Public Participation processes. In: Proceedings of the Conference on Environmental Governance and Democracy (10). Yale, Yale University Press.

Hancock, J. (2003). Environmental Human Rights: Power, Ethics and Law. London, Ashgate Publishing Limited.

Handl, G. (1992). Human Rights and Protection of the Environment: A mildly "revisionist" View. In: A. A. Cancado Trindade (Ed.) Human Rights, Sustainable Development and the Environment (p. 117). San Jose de Brasilia.

Hansen, M.H. (1991). The Athenian Democracy in the Age of Demosthenes. Oxford, University of Oklahoma Press.

Hardin, G. (1968). The Tragedy of the Commons. Science, 1243.

Hargrone, J. L. (1972). Law Institutions and global environment. New York, Kluwer Academic Publishers.

Harlow, C. (1996). Francovich and the Problem of the Disobedient State. European Law Journal, 199.

Hartley, D. (2001). Green citizenship, Social Policy and Administration, 35, 490.

Hartmann, J. (1992). Democracy, Development and Environmental Sustainability. In: H.O. Bergesen and G. Parmann (Eds.), Green Globe Yearbook of International Cooperation on Environment and Development (p. 49). Oxford, Oxford University Press.

Hauptmann, E. (2001). Can Less Be More? Leftist Deliberative Democrats. Critique of Deliberative Democracy. Policy, 397.

Hawken, P., Lovins, A. & Lovins, L. H. (2000). Natural Capitalism: Creating the next Industrial Revolution. Boston, Back Bay Books.

Hay, P. (2002). A companion to Environmental Thought. Edinburgh, Edinburgh University Press.

Hayton, R. D. (1993). The Matter of Public Participation. Natural Resources Journal. 275.

Hayward, T. (2000). Constitutional environmental rights: a case for political analysis. Political Studies, 48, 558.

Hayward, T. (2005). Constitutional environmental rights. Oxford, Oxford University Press.

Hectors, K. (2008). The Chartering of Environmental Protection: Exploring the Boundaries of Environmental Protection as Human Rights. European Energy and Environmental Law Review, June, 165.

Heilbroner, R. (1974). An Inquiry into the Human Prospect. New York, W.W. Norton.

Heinelt H., & Opladen A. (2002). Participatory Governance in Multi-Level Context: Concept and Experience. Leske + Budrich, Opladen.

Heinelt, H. (2007). Participatory Governance and European Democracy. In: B. Kohler-Kock, & B. Rittberger (Eds.), Debating the Democratic Legitimacy of the European Union (p. 219). Plymouth, Rowman & Littlefield.

Held, D. (1991). Political Theory Today. Cambridge, Stanford University Press.

Heyward, C. (2008). Can the all-affected principle include future persons? Green deliberative democracy and the non-identity problem. Environmental Politics, 17, n 4, August, 625.

Hilson, C. (2001). Greening citizenship: Boundaries of membership and the Environment. Journal of Environmental Law, 335.

Hobbes, T. (1974). Leviathan. Harmondsworth, Penguin Classics.

Hobson, K (2001). Sustainable lifestyles: rethinking barriers and behaviour change. In: M. Cohen, & J. Murphy (Eds.), Exploring sustainable consumption:

environmental policy and the social sciences (p. 191) Oxford, Emerald Group Publishing Limited.

Hobson, K. (2003). Thinking habits into action: the role of knowledge and process in questioning household consumption practices. Local Environment, 8 (1), 95.

Hodkova, I. (1991). Is There a Right to a Healthy Environment in the International Legal Order?. Connecticut Journal of International Law. 65.

Holder, J., & Lee M. (2007). Environmental Protection, Law and Policy. Cambridge, Aspen Publishers.

Holtz, U. (2008). Representative and participatory democracy. Colloquy of the European Association in Athens 02 May, Available at http://www.vemdb.de/files/rd_pd.pdf

Horn, L. (2004). The implications of the Concept of Common Concern of a Human Kind on a Human Right to a Healthy Environment. Macquarie Journal of International and Comparative EnvironmentalLaw, 1, 233.

Hostetler, E.G. (1995). Promoting the Effective Implementation of Multilateral Environmental Treaties: The Role of Non-Government organisations in Strategies For Environmental Enforcement. Stanford Environmental Law Society, 279.

Howard, W. (1996). Environmental democracy: Use it or lose it. National Wildlife, 34.

Hugo, G. (1996). Environmental concerns and international migration. International Migration Review, 105.

Jackson, T. (2005). Motivating Sustainable Consumption. Sustainable Development Research Network.

Jacobs, F. (2006). The Role of the European Court of Justice in the Protection of the Environment. Journal of Environmental Law, 18, 185.

Jacobs, M. (1997). Environmental valuation, deliberative democracy and public decision-making institutions. In: J. Foster (Ed.),Valuing nature? Economics, ethics and environment (p. 211–231). London, Routledge.

Jacobs, M. (1999). Environmental democracy. In: A. Gamble, & T. Wright (Eds.), The New Social Democracy (p. 105). Oxford, Wiley .

Jacque, J.P. (1997). La protection de l'environnement au niveau européen ou régional. In: P. Kromarek (Ed.), Environnement et droits de l'homme (p. 65). UNESCO.

James, A. (1986). Sovereign statehood the basis of international society. London, Harper Collins Publishers Ltd.

Jans, J. H. (2006). Did baron von Munchhaausen ever visit Aarhus?. In: R. Macrony (Ed.), Reflection on 30 Years of EU Environmental Law (p. 447), Groningen, Europa Law Publishing.

Jans, J. H., & Vedder, H. (2008). European Environmental Law. Groningen, Europa Law Publishing.

Jasanoff, S. (1996). The dilemma of environmental democracy. Science and Technology, 13, 2.

Jayanti, S. (2009). Recognising Global Environmental Interests: A Draft Universal Standing Treaty for Environmental Degradation. Georgetown International Environmental Law Review, 1.

Jelin, E. (2000). Towards a Global Environmental Citizenship?. Citizenship Studies, 4, 47.

Jendroska, J. (2005). Aarhus Convention and Community Law: The Interplay. Journal of European Environmental and Planning Law, 12.

Joerges, C. (2006). On the Disregard for History in the Convention Process. European Law Journal, 2.

Jóhannsdóttir, A. (2007). Considerations on the Development of Environmental Law in the Light of the Concept of Sustainable Development. YM, 27.

Jóhannsdóttir, A. (2008). Miljodemokrati–offentlighedens deltagelse i beslutningsprocessen. In: M. B. Andersen, & J. Christoffersen (Eds.), Forhandlingerne ved Det. nordiske Juristmode (p. 221), Copenhagen, University of Oslo Press.

Johnson, G. S. (2009). Environmental Justice, A brief history and overview. In: F. Chioma Steady (Ed.), Environmental justice in the new millennium: global perspectives on race, ethnicity and human rights (p. 17). New York, Palgrave Macmillan.

Johnson, G.F. (2007). Discursive democracy in the transgenerational context and a precautionary turn in public reasoning. Contemporary Political Theory, 6 (1), 67.

Jonas, H. (1979). Das Prinzip Verantwortung. Frankfurt, Königshausen & Neumann.

Jordan, L. (2000). Civil Society's Role In Global Policymaking. Alliance, March 2000. Available at www.globalpolicy.org/ngos/intro/general/2003/0520role.htm

Kaelble, H. (1994). L'Europe "vécue" et l'Europe "pensée" aux XXe siècle: Les spécificités sociales de l'Europe. In: R. Girault (Ed.), Identité et conscience européennes aux XXe siècle (p. 27). Paris, Publications de la Sorbonne.

Kanie, N. & Haas, P.M. (2004). Emerging Forces in Environmental Governance. United Nations New York, Bookwell Publications.

Karakostras, I. K. (2008). Greek and European Environmental law. Athens, Bruylant.

Karassin, O. (2010). Mind the Gap: Knowledge and Need in Regulating Adaptation to Climate Change. Georgetown International Environmental Law Review, 383.

Keessen, A. (2007). Reducing the Judicial Deficit in Multilevel Environmental Regulation: the Example of Plant Protection Products. European Environmental Law Review, 26.

Kelsen, H. (1961). General Theory of Law and State. New York, Lawbook Exchange, Ltd.

Kenny, M. (1996). Paradoxes of Community. In: B. Doherty and M. De Geus (Eds.), Democracy and Green Political Thought (p. 23). London, Routledge.

Kenyon, W., Nevin, C., & Hanley N. (2003). Enhancing Environmental Decision-making Using Citizens' Juries. Local Environment, 8, 22.

Kingsbury, B. (2007). Global Environmental Governance as Administration: Implications for International Law. In: D. Bodansky, & J. Brunnée (Eds.), International Environmental Law (p. 63). Oxford, Oxford University Press.

Kiss, A. (1976). Peut-on définir le droit de l'homme à l'environnement?. Revue Juridique de L'Environnement, 9.

Kiss, A. (1990). Le droit de la conservation de l'environnement. Revue Universelle Des Droits De L'Homme, 445.

Kiss, A. (1992). An introductory note on a human right to environment. In: E. B. Weiss (Ed.), Environmental Change and International Law (p. 1992). Hong Kong, United Nations University.

Kiss, A. (2005). De la protection intégrée de l'environnement à l'intégration du droit international de l'environnement. Revue Juridique de l'Environnement, 261.

Kiss, A. (2008). Does the European Charter of Fundamental Rights and Freedoms Guarantee a Right to Environment. In: A. Postiglione (Ed.), The role of the judiciary in the implementation and enforcement of environmental law (p. 161). Brussels, Bruylant.

Kiss, A., & Shelton, D. (2007a). Guide to International Environmental Law, New York, Brill.

Kiss, A., & Shelton, D. (2000). International Environmental Law. New York, Transnational Publishers Inc.,U.S.

Kitchen, L., Milbourne, P., Marsden, T., & Bishop, K. (2002). Forestry and Environmental Democracy: The Problematic Case of the South Wales Valleys. Journal of Environmental Policy and Planning, 4, 139.

Koester, V. (2005). Review of Compliance Under the Aarhus Convention: A Rather Unique Compliance Mechanism. Journal for European Environmental Planning Law, 31.

Kohler-Kock, B. & Rittberger, B. (2007). Debating the Democratic Legitimacy of the European Union. Plymouth, Rowman & Littlefield.

Koons, J. E. (2008). Earth Jurisprudence: the Moral Value of Nature. Pace Environmental Law Review, 263.

Kotov, V., & Nikitina, E. (1995). Russia and International Environmental Cooperation. In: H.O. Bergesen, & G. Parmann (Eds.), Green Globe Yearbook of

International Cooperation on Environment and Development (p. 17). Oxford, Oxford University Press.

Krämer L. (2008). The Environment and the Ten Commandments. Journal of Environmental Law, 20, 7.

Krämer L. (2009). Environmental Justice in the European Court of Justice. In: J. Ebbesson and Okowa P. (Eds.), Environmental law and Justice in the Context (p. 195). Cambridge, Cambridge University Press.

Kravchenko, S. (2007). The Aarhus Convention and Innovations in Compliance with Multilateral Environmental Agreements. Colorado. Journal International Environmental Law and Policy, 18, 1.

Kravchenko, S., Skrylnikov, D., & Bonine, J. E. (2003). Access to justice in cases involving public participation in decision-making. In: S. Stec (Ed.), Handbook on Access to Justice under the Aarhus Convention (p. 27). Szentendre, Unites Nation Publication.

Krisch, N., & Kingsbury, B. (2006). Introduction: Global Governance and Global Administrative Law in the International Legal Order. European Journal International Law, 1.

Krutilla, K., & Reuveny, R. (2002). The Quality of Life in the Dynamics of Economic Development. Environment and Development Economics, 7, 23.

Kumar, R. (2003a). Reasonable reasons in contractualist moral arguments. Ethics, 114.

Kumar, R. (2003b). Who can be wronged?. Philosophy and Public Affairs, 31, 99.

Laavrysen, L. (2008). The European Court of Justice and the Implementation of Environmental Law. In: A. Postiglione (Ed.): The role of the judiciary in the implementation and enforcement of environmental law (p. 25). Brussels, Bruylant.

Lador, Y. (2010). Time for a Universal Declaration on Environmental Rights. Available at www.partnerships4planet.ch/en/environmental-rights.php

Larsson, M.L. (1999). The Law of Environmental Damage – Liability and Reparation. Cambridge, Brill.

Latouche, S. (2004). Survivre au développement. Paris, Mille et une nuits.

Latta, P. A. (2007). Locating Democratic Politics in Ecological Citizenship. Environmental Politics, 16, 377.

Lee, M. (2001). From Private to Public: The Multiple Faces of Environmental Liability. European Public Law, 7, 375.

Lee, M. (2002a). New Generation regulation? The case of end-of-life vehicles. European Environmental Law Review, 114.

Lee, M. (2002b). The Changing Aims of Environmental Liability. Environmental Law and Management, 14, 189.

Lee, M. (2003). Public Participation, Procedure and Democratic Deficit in EC Environmental Law. Yearbook European Environmental law, 195.

Lee, M. (2008). The Environmental Implications of the Lisbon Treaty. Environmental Law Review, 10, 131.

Lee, M., & Abbot, C. (2003). The Usual Suspects? Public Participation Under the Aarhus Convention. The Modern Law review, 80.

Leopold, A. (1949a). A Sand Country Almanac and Sketches Here and There. Oxford, Oxford University Press.

Leopold, A. (1949b). The Land Ethic. A sand County Almanac, 204.

Lessig, L. (1995). The Regulation of Social Meaning. University of Chicago Law Review, 62, 943.

Levi, L. (2006). Access to justice at the level of the EU: a practitioner's perspective. In: A. Gourtin (Ed.), The Aarhus Regulation: New Opportunities for Citizens in the EU and Beyond? Report of the ECOSPHERE Forum held in Brussels, 27 October 2006. Brussels, ECOSPHERE.

Li, Q., & Reuveny, R. (2003). Economic Globalisation and Democracy: An Empirical Analysis. British Journal of Political Science, 29.

Lietzmann, K. M., & Gary Vest, D. (1999). Environment & Security in an International Context. North Atlantic Treaty organisation, Committee on the Challenges of Modern Society, Report 232.

Lindblom, C. (1965). The Intelligence of Democracy: Decision Making Through Mutual Adjustment. New York, The Free Press.

Linklatera, A. (1998). The transformation of political community. Cambridge, Polity Press.

Locke, J. (1968). An Essay Concerning the true Original, Extent and End of Civil Government. Social Contract, p. 5

Loibl, G. (2004). The Evolving Regime on Climate Change and Sustainable Development. In: N. Schrijver, & F. Weiss (Eds.), International Law and Sustainable Development, Principles and Practice (p. 97). Leiden, Brill.

Lorenz, P. (2007). Press officer Global 2000 – Initiative "1 Million against Nuclear Power. June 26.

Louka, E. (2004). Conflicting Integration – Environmental Law of the European Union. Cambridge, Cambridge University Press.

Low, N., & Gleeson, B. (1998). Justice, society and nature. London, Routledge.

Luque, E. (2005). Researching Environmental Citizenship and its Publics. Environmental Politics, 211.

Macdonald, K. E. (2008). A Right to a healthful Environment – Humans and Habitants: Re-thinking Rights in an Age of Climate Change. European Energy and Environmental Law review, 213.

Macedo, S. (1999). Introduction. In: S. Macedo (Ed.), Deliberative politics: essays on democracy and disagreement (p. 3). New York, OUP USA.

MacGregor, S. (2004). Reading the Earth Charter: Cosmopolitan Environmental Citizenship or Light Green Politics as Usual?. Ethics, Place and Environment, 90.

MacPherson, C. B. (1977). The Life and Times of Liberal Democracy. Oxford, OUP Canada.

Macrony, R. (2006). Reflection on 30 Years of EU Environmental Law. Groningen, Europa Law Publishing.

MacRory, R. (1996). Environmental citizenship and the law: Repairing the European Road. Journal of Environmental law, 219.

Mahoney, J. (2002). Perpetual restrictions on land the problem of the future. Virginia Law Review, 88.

Majone, G. (2002). Delegation of Regulatory Powers in a Mixed Policy. European Law Journal, 3, 319.

Makuch, Z. (2004). TBT or not TBT, That is the Question: The international Trade Law Implication of European Community Gm Traceability and Labelling Legislation. European Environmental Law Review, 226.

Maljean-Dubois, S. & Lecucq, O. (2008). Le rôle du juge dans le développement du droit de l'environnement. Bruxelles, Bruylant.

Maljean-Dubois, S. & Mehdi, R. (1999). Les Nations Unies et la protection de l'environnement: la promotion d'un développement durable. Paris, Pedone.

Manin, B. (1997). The principles of representative government. Cambridge, Cambridge University Press.

Mank, B. (1996). Protecting the environment for future generations: a proposal for a republican superagency. New York University Environmental Law Journal, 5, 445.

Markey, B. (2004). The Earth Charter and Ecological Integrity – Some Policy implications. World View, 76.

Marks, S. (2004). The Human Right to Development: Between Rhetoric and Reality. Harvard Human Rights, 17, 137.

Marsden, S., & De Mulder, J. (2005). Strategic Environmental Assessment and Sustainability in Europe – how bring is the Future?. Review of European Community and International Environmental Law, 50.

Marsh, G. P. (1864). Man and Nature. New York, Kessinger Publishing.

Marshall, F. (2006). Two Years in the Life: The Pioneering Aarhus Convention Compliance Committee 2004–2006. International Community Law Review, 8, 123.

Martens, M. (2007). Constitutional Right to a Healthy Environment in Belgium. Review of European Community and International Environmental Law, 287.

Mason, M. (1999). Environmental Democracy. London, Routledge.

Mathiesen, A. (2003). Public Participation in Decision-making and Access to Justice in EC Environmental Law: the Case of Certain Plans and Programmes. European Environmental Law Review, 36.

Maurer, A. (2007). The European parliament between Policy-Making and Control. In: B. Kohler-Kock, & B. Rittberger (Eds.), Debating the Democratic Legitimacy of the European Union (75). Plymouth, Rowman & Littlefield.

May, J. R. (2005-2006). Constituting Fundamental Environmental Rights Worldwide. Pace Environmental Law Review. 113.

McAdams, R. (1997). The Origin, Development, and Regulation of Norms. Michigan Law Review, 96, 338.

McCaffrey, S. C., & Lutz, R.E. (1978). Environmental pollution and individual rights: an international symposium. London, Kluwer.

McCormick, J. (1995). The Global Environment Movement. Wiley, Wiley-Blackwell.

McManus, F. (2005). Noise Pollution and Human Rights, European Human Rights Law Review. 575.

Meadowcroft, J. (1997). Planning, democracy and the challenge of sustainable development. International Political Science Review, 2, 167.

Mega, V. P. (1997). Fragments of an Urban Discourse in Europe: Utopias and Europias. A sustainability-friendly ABC. In: V. P. Mega, & R. Petrella (Eds.), Utopias and Realities of Urban Sustainable Development, New Alliances between Economy, Environment, and Democracy for small and Medium-sized Cities (p. 47). Dublin, European Foundation for the Improvement of Living and Working Conditions.

Mega, V. P. & Petrella, R. (1997). Utopias and Realities of Urban Sustainable Development, New Alliances between Economy, Environment, and Democracy for small and Medium-sized Cities. Dublin, European Foundation for the Improvement of Living and Working Conditions.

Melle, U. (1998). Responsibility and the Crisis of Technological Civilization: A Husserlian Meditation on Hans Jonas. Human Studies, 21, 329.

Melo-Escrihuela, C. (2008). Promoting Ecological Citizenship: Rights, Duties and Political Agency. An International E-Journal for Critical Geographies, 113.

Mendez, A. J. (2005). Between Laeken and the Deep Blue Sea. An Assessment of the Draft Constitutional Treaty from a Deliberative-Democratic Standpoint. European Public Law, 105.

Merchant, C. (2005). Radical Ecology: The Search for a Livable World. London, Routledge.

Mersel, Y. (2006) The dissolution of political parties: the problem of internal democracy. International Journal of Constitutional law, 84.

Midlarsky ,M. (1998). Democracy and the Environment: An Empirical Assessment. Journal of Peace Research, 35, 341.

Milton, K. & Curtin, D. (2002). Ecological Citizenship. In: F. Engin, & S. Turner (Eds.), Handbook of Citizenship Studies (293). London, , SAGE Publications Ltd..

Minteer, B. A., & B. Pepperman Taylor. Democracy and the Claims of Nature: Critical Perspectives for a New Century. Oxford, Rowman & Littlefield Publishers Inc..

Mollo, M. (2005). Environmental Rights Report: Human Rights and the Environment. (Materials for the 61st Session of the United Nations Commission on Human Rights, Geneva, March 14-April 22,), online: Earthjustice Legal Defense Fund.

Montanari, P. & Corradini, A. (2008). Women and Environment: women's sensibility in Innovative environmental projects. In: A. Postiglione (Ed.), The Protection and Sustainable Development of the Mediterranean Black Sea Ecosystem (p. 863). Bruxelles, Bruylant.

Montaro, R. (2002). L'ambiente e i nuovi istituti della partecipazione. In: A. Crosetti, & F. Fracchia, Procedimento amministrativo e partecipazione. Problemi, prospettive ed esperienze (p. 114). Milano, Giuffré.

Montefiore, H. (1970). Can Man Survive?. London, Fontana.

Morgera, E. (2005). An Update on the Aarhus Convention and its Continued Global Relevance. Review of European Community and International Environmental Law, 138.

Mori, S. (2004). Institutionalization of NGO involvement in policy functions for global environmental governance. In: N. Kanie, & P.M. Haas (Eds.), Emerging Forces in Environmental Governance (157). United Nations New York.

Morrison, R. (1995). Ecological Democracy. Boston.

Mularoni, A. (2008). The Right to a safe environment in the case-Law of the European Court of human rights. In: Postiglione A. (Ed.), The role of the judiciary in the implementation and enforcement of environmental law (p. 231). Brussels, Bruylant.

Mullerat, B. (2005). European Environmental Liability: One Step Forward. International Company and Commercial Law Review, 16, 263.

Murphi, J., & Cohen, M. (2001). Sustainable consumption: environmental policy and the social sciences. In: M. Cohen, & J. Murphi (Eds.), Exploring sustainable consumption: environmental policy and the social sciences (p. 225). Oxford, Emerald Group Publishing Limited.

Myers, N. (1993). Environmental refugees in a globally warmed world. Bioscience, 43, 752.

Myers, N. (1997). Environmental refugees. Population and Environment, 19, 167.

Myint, T. (2003). Democracy in Global Environmental Governance: Issues, Interests, and Actors in the Mekong and the Rhine. Indiana Journal of Global Legal Studies, 10, 287.

Nadal, C. (2008). Pursuing Substantive Environmental Justice: The Aarhus Convention as a Pillar of Empowerment. Environmental Law Review, 10, 28.

Naess, A. (1956). Democracy, Ideology and Objectivity. Oslo, Published for the Norwegian Research Council for Science and the Humanities by Oslo U.P; Blackwell.

Naess, A. (1990). Ecology, Community, and Lifestyle. Cambridge, Cambridge University Press.

Nagel, T. (1986). The view from nowhere. New York, OUP USA.

Najam, D., Papa, M. & Taiyab, N. (2006). Global Environmental Governance: A Reform Agenda. Winnipeg, International Institute for Sutainable Development (IISD).

Nanda, P., & Pring, G. (2003). International Environmental law for the 21st Century. New York, Transnational Publishers Inc.

Nascimento, A. (2009). Global frameworks for environmental justice: Searching for global responses to global problems, 8th Global Conference "Environmental Justice and Global Citizenship. 10th – 12th July Oxford, Available at www.inter-disciplinary.net/critical-issues/ethos/environmental-justice-and-global-citizenship /project-archives/8th/.

Nelson, J. (2002). Building Partnerships: Cooperation between the United Nations System and the Private Sector. New York, United Nations.

Neumayer, E. (2002). Do Democracies Exhibit Stronger International Cross Sectional Analysis. Journal of Peace Research, 139.

Newigl, J., & Fritsch O (2009). Environmental Governance: Participatory, Multi-Level – and Effective?. Environmental Policy and Governance, 19, 197.

Nickel, J. W. (1993). The Human Right to a Safe Environment: Philosophical Perspectives on Its Scope and Justification. Yale Journal International Law, 18, 281.

Nicolet,C. (1978). Le métier de citoyen dans la Rome antique. Paris, Gallimard.

Nino, C. S. (1996). The Constitution of Deliberative Democracy, Oxford, OUP Oxford.

Norton, B. (2000). Biodiversity and Environment Values: In Search of a Universal Earth Ethic. Biodiversity and Conservation, 1029.

Noss, R. F. (1994). Some Principles of Conservation Biology, as They Apply to Environmental Law. Chicago Law Review, 893.

O'Neill, J. (2002). Deliberative Democracy and Environmental Policy. In: B. A. Minteer & B. Pepperman Taylor (Eds.), Democracy and the Claims of Nature (p. 257). Oxford, Rowman & Littlefield Publishers Inc..

Offe, C. & Preuss, U. K. (1991). Democratic institutions and Moral Resources. In: D. Held (Ed.), Political Theory Today (p. 165). Cambridge, Stanford University Press.

Ollennu, N.A. (1962). Principles of Customary Land Law in Ghana. London, Sweet & Maxwell.

Ophuls, W. (1973). Leviathan or Oblivion?. In: H. E. Daly (Ed.), Toward a Steady State Economy (p. 224). San Francisco, W.H. Freeman & Co Ltd..

Ophuls, W. (1977). Ecology and the Politics of Scarcity: A Prologue to a Political Theory of the Steady State. San Francisco, W.H.Freeman & Co Ltd.

Ostrogorski, M. (1992). Democracy and the organisation of political parties. New York, Forgotten Books.

Pace, V. (2001). La comunità religiosa internazionale e l'ambiente. In: A. Postiglione, & A. Pavan,(Eds.) Etica Ambiente Sviluppo (p. 15). Napoli, Edizioni Scientifiche Italiane.

Page, E. (2007). Climate change, justice and future generations. Cheltenham, Edward Elgar Publishing Ltd.

Pallemaerts, M. (2002). Introduction: human rights and environmental protection. In: M. Déjeant-Pons, & M. Pallemaerts (Eds.), Human Rights and the environment (p. 11). Strasbourg, Council of Europe..

Pallemaerts, M. (2003a). Human rights and democracy in the face of international environmental issues, available at www.coe.int/T/E/Com/Press/colloquies/2003/Pallemaerts_report.asp.

Pallemaerts, M. (2003b). Is Multilateralism the Future? Sustainable Development or Globalisation as 'A Comprehensive Vision of the Future of Humanity. Environment, Development and Sustainability, 275.

Pallemaerts, M. (2004). Proceduralising environmental rights: the Aarhus Convention on Access to Information, Public Participation in Decision-Making and Access to Justice in Environmental Matters in a Human Rights Context. In: Human Rights and the Environment Proceedings of a Geneva Environment Network roundtable (p.14). Geneve, Geneva Environment Network.

Pallemaerts, M. (2009). Compliance by the European Community with its obligations on Access to Justice as a Party to the Aarhus Convention. London, IEEP Report (Institute for European Environmental Policy).

Pallemaerts, M., & Moreau, M. (2004). The role of « stakeholders » in international environmental governance. Global Governance, 15.

Panayoto, T. (2000). Economic Growth and the Environment. Cambridge. University of Cambridge Press.

Paper, E. & Redclift, M. (2003). Human Security and the Environment: International Comparisons. London, Edward Elgar Publishing.

Parfit, D. (1987). Reasons and persons. Oxford, Oxford University Press.

Park, J., Conca, K., & Finger M. (2008). The crisis of Global Environmental Governance. London, Routledge.

Park, J., Conca, K., & Finger, M. (2008). The death of Rio environmentalism. In: J. Park, K. Conca, & M. Finger (Eds.), The crisis of Global Environmental Governance (p. 1). London, Routledge.

Parker, K. (1992). State Liability in Damages for Breach of Community Law. Law Quarterly Review, 181.

Parry, J. (2010). Participatory democracy in the EU. Available at www.federalunion. org.uk/europe/participatorydemocracy.shtml.

Pasques, M. (2006). L'Environnement, un certain droit de l'homme. Administration Publique, 40.

Passmore, J. (1975). Attitude to Nature. Nature and Conduct, 251.

Passmore, J. (1974). Man's Responsibility for Nature: Ecological Problems and Western Traditions. London, Macmillan Pub Co.

Pateman, C. (1970). For a democratic polity to exist it is necessary for a participatory society to exist. In: C. Pateman (Ed.), Participation and Democratic Theory (p. 43). Cambridge, Cambridge University Press.

Pateman, C. (1970). Participation and Democratic Theory. Cambridge, Cambridge University Press.

Pathak, R.S. (1992). The Human Rights System as a Conceptual Framework for Environmental Law. In E. B. Weiss (Ed.), Environmental Change and International Law: New Challenges and Dimensions (p. 205). Hong Kong, United Nations University.

Paul, J. A. (2000). NGOs and Global Policy-Making. Global Policy Forum, June 2000. Available at www.globalpolicy.org/ngos/analysis/anal00.htm

Payne, R. A. (1995). Freedom and the Environment. Journal of Democracy, 6, 41.

Pedersen, O. W. (2010). European Environmental Human Rights and Environmental Rights: A Long Time Coming?. Available at www.ssrn.com/abstract=1122289

Peel, J. (2001.) Giving the public a voice in the protection of the global environment: avenues for participation by NGOs in dispute resolution at the European Court of Justice and World Trade Organisation. Colorado Journal International Law Environmental Law and Policy, p. 47

Pennera, C., & Schoo, J. (2004). La Codécision–dix ans d'application. Cahiers de droit européen, 531.

Pennock, J.R. & Chapman, J.W. (1983). Liberal democracy. New York, New York University Press.

Perrez, F. X. (2008). How to Get Beyond the Pareto Optimum of Stakeholder Participation in Environmental Governance. In: Proceedings of the Conference on Environmental Governance and Democracy. Yale, Yale University Press.

Petit, B. (2004). La dimension sociale du développement durable: le parent pauvre du concept. Les Petites Affiches, n° 12, 8.

Petit, Y. (2009). Droit et politiques de l'environnent. Paris, La documentation Française.

Petkova, E., & Veit, P. (2000). Environmental Accountability Beyond the Nation – State: The implications of the Aarhus Convention. Washington, World Resources Institute.

Pevato, P. (1989). A Right to Environment in International Law: Current Status and Future Outlook. Review of European Community and International Law, 8, 309.

Pieratti, G., & Prat, J.-L. (2000). Droit, économie, écologie et développement durable: des relations nécessairement complémentaires mais inévitablement ambiguës. Revue Juridique de l'Environnement, 422.

Pildes, R. H., & Niemi, R.G. (1993). Expressive Harms, "Bizarre Districts," and Voting Rights: Evaluating Election-District Appearances After Shaw v. Ren. Michigan Law Review, 92, 438.

Plumwood, V. (1999). Inequality, ecojustice and ecological rationality. Ecotheology, 5/6, 185.

Posner, E. A. (2000). Law and Social Norms: The Case of Tax Compliance. Virginia Law Review, 86, 1781.

Postiglione, A. (2008). The role of the judiciary in the implementation and enforcement of environmental law. Brussels, Bruylant.

Poujade, B. (2007). La Protection du droit à l'environnement par le juge administratif: l'exemple du contrôle juridictionnel des grands équipements public. In. A. Chamboredon (Ed.), Du Droit de l'Environnement au droit à l'Environnement, A la recherche d'un juste milieu (p.123). Paris, L'Harmattan.

Powell, F.M. (1995). Environmental Protection in International Trade Agreements: The Role of Public Participation in the Aftermath of NAFTA. Colorado Journal International Law Environmental Law and Policy, 109.

Prieur, M. (1993). Démocratie et droit de l'environnement et du développement. Revue Juridique de l'Environnement, p. 23.

Prieur, M. (2005). Les nouveaux droits. Actualité Juridique Droit Administratif, 1157.

Ramlogan, R. (1996). Environmental refugees. Review Environmental Conservation, 23, 81.

Raustiala, K. (1997). The Participatory Revolution in International Environmental Law. Harvard Journal International Law Environmental Law and Policy, 537.

Rawls, J. (1971). A theory of justice. Oxford, Harvard University Press.

Rawls, J. (1993). Political liberalism. New York, Columbia University Press.

Razaque, J. (2002). Background Paper Number 4: Human Rights and the Environment: The National Experience of South Asia and Africa, 2002. Available at www.unhchr.ch/environment/bp4.html.

Redgwell, C. (2007). Access to Environmental Justice. In: F. Francioni (Ed.), Access to Justice as a Human Right (p. 153). Oxford, Oxford University Press.

Reiman, J. (2007). Being fair to future people: the non-identity problem in the original position. Philosophy and Public Affairs, 35, 69.

Reiners, K. (2009). The Environmental Liability Directive of 2004, Traditional Administrative Mechanisms with a New Name, LL.M. Thesis. Reykjavik.

Remond- Gouilloud, M. (1982), La „Charte de la nature". Revue juridique de l'environnement, 120.

Renn, O. (1995). Fairness and competence in citizen participation: Evaluation models for environmental discourse, Dordrecht, Springer.

Rensi, G. (1995). La Democrazia Diretta. Milano, Adelphi.

Rest, A. (2008). Access to justice in international environmental law for individuals an NGOs: Efficacious enforcement by the Permanent Court of Arbitration. In: A. Postiglione (Ed.), The role of the judiciary in the implementation and enforcement of environmental law (p. 56). Bruxelles, Bruylant.

Reuveny, R. (2003). Economic Growth, Environmental Scarcity and Conflict. Global Environmental Politics, 83.

Richardson, B. J. & Razzaque, J. (2006). Public Participation in Environmental Decision-making. In: B. J. Richardson, & S. Wood (Eds.), Environmental Law for Sustainability (p. 166). Oxford, Hart Publishing.

Richardson, B. J. & Wood, S. (2006). Environmental Law for Sustainability. Oxford, Hart Publishing.

Roberts, A. (2001). Structural Pluralism and the Right to Information. University of Toronto Law Journal, 243.

Robinson, N. A. (1972). Problems of definition and scope. In: J. L. Hargrone, (Ed.), Law Institutions and global environment (p. 44). New York, Kluwer Academic Publishers.

Robinson, N. A. (2003). Enforcing Environmental Norms: Diplomatic and Judicial Approaches. Hastings International and Comparative Law Review, 387.

Rocheleau, J. (1999). Democracy and Ecological Soundness. Ethics and the Environmental, 4, 38.

Rodenhoff, V. (2002). The Aarhus Convention and its Implications for the 'Institutions' of the European Community. Review of European Community and International Environmental Law, 11, 343.

Rodgers, W.H. (1977). Environmental Law. London, West Publishing Co.

Rodriguez, X. (2006). The Water Framework directive and the polluter pay principle, LLM Thesis 2006, University of Iceland.

Rodriquez, S. (2008). Representative democracy Vs. Participatory democracy in the EU and the US. In: R. Caranta (Ed.), Interest representation in administrative proceeding (p. 24). Napoli, Jovene.

Rolson, H. (1988). Environmental Ethics: Duties to and Values in the Natural World. Philadelphia, 143.

Rolston, H. (1993). Rights and Responsibilities on the Home Planet. Yale Journal International Law, 251.

Romi, R., Bossis, G., & Rousseau, S. (2005). Droit international et européen de l'environnement, Paris, Montchrestien.

Rose-Ackerman, S., & Halpaap, A. A. (2001). The Aarhus Convention and the Politics of Process: The Political Economy of Procedural Environmental Rights, Draft paper for The Law and Economics of Environmental Policy: A Symposium, Faculty of Laws, University College London, September 5-7, 2001. Available at www.cserge.ucl.ac.uk/Ackerman_and_Halpaap.pdf.

Rosenne, S. (1986). Exploitation and Protection of the Exclusive Economic Zone and the Continental Shelf. Yacht Brokers Association of America, 63.

Ross, M. (1993). Beyond Francovich. Modern Law Review, 56, 55.

Roston, H. (1993). Rights and Responsibilities on the Home Planet. Yale Journal International Law, 251.

Rousseau, J. J. (1913). The Social Contract. Oxford, Oxford Paperbacks.

Rousseau, J. J. (1992). Du contrat social. Paris, Bréal.

Ruster, B., & Simma, B. (1990). International Protection of the Environment. New York, Oceana Pubns.

Ryall, A. (2007). EIA and Public participation: Determine the limits of Members State Discretion, Case law analysis, Case C-216/05, Commission v. Ireland, judgement of 9 November 2006. Journal of Environmental Law, 247.

Sáiz, A.V. (2005). Globalisation, Cosmopolitanism and Ecological Citizenship. Environmental Politics, 14(2), 163.

Sanchez, R. (1993). Public Participation and the IBWC: Challenges and Options. Natural Resources Journal, 283.

Sands, P. (1991). The Role of Non-Governmental Organisations in Enforcing International Environmental Law. In: W. E. Butler (Ed.), Control Over Compliance With International Law. London, Kluwer Academic Publishers.

Sands, P. (1995). Principle of International Environmental Law. Cambridge, Cambridge University Press.

Sands, P. (1999). Sustainable Development: Treaty, Custom, and the Cross-Fertilisation of International Law. In: A. Boyle, & D. Freestone (Eds.), International Law and Sustainable Development (p. 43). Oxford, Brill.

Sands, P. (2004). Human rights and the environment. In M. Pallermarts, Human Rights and the Environment Proceedings of a Geneva Environment Network roundtable. Geneva, Geneva Environment Network.

Savoia, R. (2003). Administrative, judicial and other means of access to justice. In: S. Stec (Ed.), Handbook on Access to Justice under the Aarhus Convention (p. 39), Szentendre, Unites Nation Publication.

Saward, M. (1998). Green state/democratic state. Contemporary Politics, 4, 345.

Saward, M. (2001). Reconstructing democracy: Current thinking and new directions. Government and Opposition, 36, 559.

Sax, J. L. (1972). Defending the Environment, A Handbook for Citizen Action. New York, Vintage Books.

Scanlon, T.M. (1999). What we owe to each other. Cambridge, Belknap Pr.

Schachter, O. (1997). The Decline of the Nation-State and its Implications for International Law. In: J. I. Charney, D. K. Anton, & M. E. O'Connell (Eds.), Politics, Values And Functions: International Law In The 21st Century: Essays In Honor Of Professor Louis (p. 26). Henkin, Martinus Nijhoff Publishers.

Schall, C. (2008). Public Interest Litigation Concerning Environmental Matters before Human Rights Courts: A Promising Future Concept?. Journal of Environmental Law, 417.

Schechter, M.G. (2001). Making meaningful UN-sponsored world conferences of the 1990s: NGOs to the rescue?. In: M.G. Schechter (Ed.), United Nations-Sponsored World Conferences: Focus on Impact and Follow-up (p. 184). Tokyo, United Nations.

Schlosberg, D., Shulman, S.,& Zavetoski, S. (2006). Virtual environmental citizenship: Web-based public participation in rule making in the United States. In: A. Dobson, & D. Bell (Eds.), Environmental Citizenship (p. 207). Cambridge, Cambridge University Press.

Schmalz-Bruns, R. (2002). Normative Desirability of Participatory Democracy. In: H. Heinelt, & A. Opladen (Eds.), Participatory Governance in Multi-Level Context: Concept and Experience (p. 59). Leverkusen, Leske + Budrich.

Schmalz-Bruns, R. (2007). The Euro-polity in Perspective: Some Normative Lessons from Deliberative Democracy, In: B. Kohler-Kock & B. Rittberger (Eds.), Debating the Democratic Legitimacy of the European Union (p. 281). Plymouth, Rowman & Littlefield.

Schmitt, C. (1999). Parlamentarismo e democrazia, Lungro, Marco.

Schmitter, P. C., & Lehmbruch G. (1979). Trends Toward Corporatist Intermediation. Sage, SAGE Publications Ltd.

Schrijve, N. (2008). The evolution of sustainable development in international law: inception, meaning and status. Boston, The Hague Academy of International Law.

Schrijver, N. (1995). Sovereignty Over Natural Resources – Balancing Rights and Duties in an Independent World. Groningen, Europa Law Publishing.

Schrijver, N. & Weiss, F. (2004). International Law and Sustainable Development, Principles and Practice. Leiden, Brill.

Schultz, C. B., & Crockett, T.R. (1990). Economic Development, Democratisation, and Environmental Protection in Eastern Europe. Boston College Environmental Affairs Law Review, 18, 53.

Sedjari, I A. (2008). Droits de l'homme et développement durable. Paris, L'Harmattan.

Sen, A. (1994). Liberty and Poverty: Political Rights and Economics. New Republic, 31.

Sensi, S. (2004). Human Rights and the Environment: The Perspective of the Human Rights Bodies. In UN Environment Programme, Human Rights and the Environment, Proceedings of a Geneva Environmental Network Roundtable (UNEP, 2004). Available at www.environmenthouse.ch/docspublications/ reportsRoundtables/Human%20/Rights%20Env%20Report.pdf>.

Seyfang, G. (2005). Shopping for Sustainability: Can Sustainable Consumption Promote Ecological Citizenship?. Environmental Politics, 14, 290.

Shama, S. (1995). Landscape and memory. Fontana, Harper Perennial.

Sharander-Frechette, K. (2002). Environmental justice: creating equality, reclaiming democracy. Oxford, OUP USA.

Shelton, D. (1991). Human Rights, Environmental Rights and the Right to Environment. Stanford Journal of International Law, 28, 103.

Shelton, D. (1992). What happened in Rio to human rights?. Yearbook International Environmental Law, 82.

Shelton, D. (2005). Environmental Rights. In: P. Alston (Ed.), People's Rights (p. 185). Oxford, OUP Oxford.

Shelton, D. (2007). Human rights and the environment: what specific environmental rights have been recognised?. Denver Journal of International Law and Policy, 35, 129.

Shelton, D., & Memon, A. (2002). Adopting Sustainability as an Overarching Environmental Policy: A review of Section 5 of the RMA. Resource Management Journal, March, 8.

Sieghart, P. (1985). The Lawful Rights of Mankind: An introduction to the International Legal Code of Human Rights. Oxford, Oxford University Press.

Sjafjell, B. (2010). The Very Basis of Our Existence Labour and the Neglected Environmental Dimension of Sustainable Development. Available at www.ssrn.com/abstract=1517393.

Skagen, K. (2005). Giving a voice to posterity – deliberative democracy and representation of future people. Journal of Agricultural and Environmental Ethics, 18, 429.

Skinner, J.B. (1988). Earth Resources. New Jersey, Prentice Hall.

Smith, G. (2003). Deliberative democracy and the environment. London, Routledge.

Smith, G. (2005). Green Citizenship and the Social Economy. Environmental Politics, 14, 273.

Smith, G., & Wales, C. (2000). Citizens' juries and deliberative democracy. Political Studies, 1, 51.

Smith, M.J. (1998). Ecologisme–Towards Ecological Citizenship. London, Open University Press.

Snyder, F. (2004). Editorial: Is the European Constitution Dead?. European Law Journal, 255.

Sohn, L. (1973). The Stockholm Declaration on the Human Environment. Harvard International Law Journal, 455.

Solum, L. (2001). To our children's children: The problem of intergenerational ethics. Loyola of Los Angeles law Review, 35.

Soveroski, M. (2007). Environment Rights versus Environmental Wrongs: Forum over Substance?. Review of European Community and International Environmental Law, 16, 261.

Spaulding, M. J. (1995). Transparency of Environmental Regulation and Public Participation in the Resolution of International Environmental Disputes. Santa Clara Law Review, 1127.

Speeckaert, G. P. (1956). Les fonctions, les méthodes et la valeur du travail international non gouvernemental. In: L'avenir des organisations non gouvernementales (p. 39). Brussels, Union des Associations internationales.

Speth, J.G. (2009). The Bridge at the Edge of the World: Capitalism, the Environment, and Crossing from Crisis to Sustainability. London, Yale University Press.

Spyke, N.P. (1999). Public Participation in Environmental Decision-making at the New Millennium: Structuring New spheres of Public Influence. Boston College Environmental Affairs Law Review, 26, 263.

Stange, T., & Baylet, A. (2008). Le développement durable à la croisée de l'économie, de la société et de l'environnement. Paris, La Documentation Française.

Staveley, E.S. (1972). Greek and Roman Voting and Elections. New York, Cornell Univ Pr.

Stec, S. (2003). Handbook on Access to Justice under the Aarhus Convention. Szentendre, Unites Nation Publication.

Stec, S. & Casey-Lefkowitz S. (2000). The Aarhus Convention: An Implementation Guide. New York and Geneva, United Nations.

Steele, J. (2001). Participation and Deliberation in Environmental Law: Exploring a problem-solving approach. Oxford Journal of Legal Studies, 21, 415.

Steffek, J., & Nanz, P. (2005). Deliberation and Democracy in Global Governance: The Role of Civic Society. In: S. Thoyer, & B. Martimort-Asso, (Eds.), Participation for Sustainability in Trade. London, Ashgate.

Stein, J P. (1995). A Specialist Environmental Court: An Australian Experience. In D. Robinson, & J. Dunkley. (Eds.), Public Interest Perspectives in Environment Law. London, John Wiley & Sons.

Stein, T. (1998). The Ecological Crisis – A Challenge to Constitutional Democracy? Does the Constitutional and Democratic System Work?. Constellations, 4, 420.

Stephenson, W. (1978). Concourse theory of communication. Communication, 3, 21.

Stoczkiewciz, M. (2009). The polluter pay principle and State aid for environmental protection. Journal of European Environmental and Planning Law, 171.

Stone, C. (1972). Should threes have Standing? Towards Legal Rights for Natural Objects. S. Cal. Law Review, 450.

Stone, C. (1988). The Environment in Moral Thought. Tennessee Law Review, 56.

Stone, C. (2007). Ethics and International Environmental Law. In: D. Bodansky, & J. Brunnée (Eds.), International Environmental Law (p. 291). Oxford, Oxford University Press.

Sudre, F. (2001). Droit International et Européen des Droits de L'Homme. Paris, Pr. Univ. de France.

Suhrke, A. (1994). Environmental degradation and population flows. Journal of International Affairs, 47, 473.

Sverker, C. J. (2009). In search of the ecological citizen. Environmental Politics, 18, 18.

Takacs, D. (2010). Forest Carbon Offsets and International Law: A Deep Equity Legal Analysis. Georgetown International Environmental Law Review, 521.

Tan Yan, Wang Yi Qian, (2004). Environmental Migration and Sustainable Development in the Upper Reaches of the Yangtze River. Population and Environment, 25, 613.

Taylor, M. (1982). Community, Anarchy and Liberty. Cambridge, Cambridge University Press.

Taylor, P. (1998). From Environmental to Ecological Human rights. A new Dynamic in International Law. Geo International Environmental Law Review, 10, 309.

Taylor, P. (2009). Ecological Integrity and Human Rights. In: L. Westra, K. Bosselmann, & R. Westra (Eds.), Reconciling Human Existence with Ecological Integrity (p. 89). London, Earthscan.

Taylor, P. W. (1986). Respect for nature: a theory of environmental ethics. Princeton, Princeton University Press.

Teubner, G., & Other (1994). Environmental Law and Ecological Responsibility: The concept and Practice of Ecological Self-Organisation, Chichester, Wiley-Blackwell.

Thieffry, P. (2008). Droit de l'environnement de l'Union Européenne. Bruxelles, Bruylant.

Thomas, D.S.G., & Twyman, C. (2005). Equity and justice in climate change adaptation amongst natural-resource-dependent societies. Global Environmental Change, 15, 115.

Thompson, B. (2004). The trouble with time: Influencing the Conservation Choices of Future Generations. Natural Resources Journal, 44.

Thompson, D. (2005). Democracy in time: popular sovereignty and temporal representation. Constellations, 12, 245.

Thornton, J., & Beckwith, S. (2004). Environmental Law. London, Sweet & Maxwell.

Thoyer, S. & Martimort-Asso, B. (2005). Participation fort Sustainability in Trade. London, Ashgate.

Torgerson, D. (2008). Constituting Green Democracy: A political project. The Good Society, 17, 18.

Torras, M., & Boyce J. K. (1998). Income, Inequality, and Pollution: A Reassessment of the Environmental Kuznets Curve. Ecological Economy, 147.

Toth, A. G. (1992). The Principle of Subsidiary in the Maastricht Treaty. Common Market Law Review, 239.

Touzet, A. (2008). Droit et développement durable. Revue Droit Public, 453.

Tribe, L. H. (1974). Ways To Think About Plastic Trees: New Foundations for Environmental Law. The Yale Law Journal, 1315.

Tromans, S. (2001). Environmental Court Project: Final Report. Journal of Environment Law, 13, 423.

Tuesen, J., & Simonsen, J. (2000). Compliance with the Aarhus Convention. Environmental Policy and Law, 299.

Tully, J. (2002). The freedom of the moderns in comparison to their ideas of constitutional democracy. Modern Law Review, 65, 204.

Turner, S. J. (2008). A Substantive Environmental Right. Austin. Kluwer Law International.

UNEP, (1991). Caring for the Earth: A Strategy for Sustainable Living, Gland, (Switzerland), IUCN, UNEP and WWF.

Valdivielso, J. (2005). Social Citizenship and the Environment. Environmental Politics, 14, 239.

Van Lang, A. (2007). La protection constitutionnelle du droit à l'environnement. In: A. Chamboredon (Ed.), Du Droit de l'Environnement au droit à l'Environnement, A la recherche d'un juste milieu (p. 123). Paris, L'Harmattan.

Verschuuren, J. (2004). Public Participation regarding the Elaboration and Approval of Projects in the EU after the Aarhus Convention. Yearbook European Environmental law, 35.

Visalenc, G. (2008). Le développement durable: regards sur un droit en construction, et sur ses bâtisseurs. Les Petites Affiches, 22 avril, n° 81, 8.

Vitale, D. (2006). Between deliberative and participatory democracy: A contribution on Habermas. Philosophy Social Criticism, 32, 748.

Von Bogdandy, A. (2005). The Prospect of a European Republic: what European Citizens are voting on. Common Market Law Review, 913.

Vonkeman, G. (1997). Alliances between Economy, Ecology and Democracy: Integration or Separation? Proposals and Analysis on the Relations between the Three Spheres; Repercussions for the Small and Medium-Sized Cities of Europe. In: V. P. Mega & R. Petrella (Eds.), Utopias and Realities of Urban Sustainable Development, New Alliances between Economy, Environment, and Democracy for small and Medium-sized Cities (p. 319). Dublin, European Foundation for the Improvement of Living and Working Conditions.

Waak, P. (1995). Shaping a Sustainable Planet: The Role of Nongovernmental Organisations. Colorado Journal International Law Environmental Law and Policy, 345.

Waite, A. (2007). Sunlight through the trees, a perspective on environmental rights and human rights. In: Mélanges en l'Honneur de Michel Prieur, Pour un droit commun de l'environnement (p. 393), Paris, Dalloz.

Wälde, T. (2004). Natural Resources and Sustainable Development: From "Good Intentions" of "Good Consequences". In: N. Schrijner & F. Weiss (Eds.), International Law and Sustainable Development, Principles and Practice (p. 119). Leiden, Brill.

Walker, K.J. (1988). The Environmental Crisis: A Critique of Neo-Hobbesian Responses. Polity, 21, 67.

Wang, X., & Wart, M.W. (2007). When Public Participation in Administration Leads to Trust: An empirical Assessment of Managers' Perception. Public Administration Review, 265.

Warren, M. (1995). The self in Discursive Democracy. In: S. While (Ed.), The Cambridge Companion to Habermas (p. 181). Cambridge, Cambridge University Press.

Wates, J. (2005a). The Aarhus Convention: a Driving Force for Environmental Democracy. Journal of European Environmental Planning Law, 2.

Wates, J. (2005b). The Aarhus Convention: a new instrument Promoting Environmental Democracy. In: M.-C Cordonier Segger, & C.G. Weeramantry (Eds.), Sustainable Justice: Reconciling Economic, Social and Environmental Law (p. 393). London, Martinus Nijhoff.

Webler, T., & Renn, O. (1995). A brief primer on participation: philosophy and practice. Fairness and Competence in Citizen Participation: Evaluating Models for Environmental Discourse. Netherlands, Springer.

Weiler, J. (1997). Legitimacy and Democracy of Union Governance. In: G. Edwards, & A. Pijpers (Eds.), The Politics of European Treaty Reform (p. 249). London, Continuum International Publishing Group Ltd.

Weinburg, R. (2010). Review: future people: a moderate consequentialist account of our obligations to future generations. Notre Dame Philosophical Reviews Notre Dame University. Available at www.ndpr.nd.edu/review.cfm?id1/48165

Weiss, E. B. (1984). Conservation and Equity between Generations. Contemporary Issues in International Law, 119.

Weiss, E. B. (1989). In Fairness to Future Generation: international Law, common patrimony and Intergeneration Equity. Tokyo, Transnational Publishers Inc.

Weiss, E. B. (1990). Our rights and obligations to future generations for the environment. American Journal of International Law, 198.

Weiss, E. B. (1992). Intergenerational equity: A legal framework for global environmental change. In: E. B. Weiss (Ed.), Environmental Change and International Law. Tokyo, United Nations University.

Weiss, E. B. (2000). The Rise or the Fall of International Law?. Fordham Law Review, 345.

Wenneras, P. (2007). The Enforcement of EC Environmental law. Oxford, OUP Oxford.

Westerlund, E. (2008). A Sustainable Criminal Law–Criminal law for Sustainability. In: H. C. Bugge, & C. Voigt (Eds.), Sustainable Development in International and National Law (p. 503). Groningen, Europa Law Publishing.

Westing, A.H. (1992). Environmental refugees: A growing category of displaced persons. Environmental Conservation, 19, 20.

Westing, A.H. (1994). Population, desertification and migration. Environmental Conservation, 21, 110.

Westra, L. (2006). Environmental Justice and the Rights of Unborn and Future Generations. London, Routledge.

Westra, L. (2004). Ecoviolence and the Law, Supranational Normative Foundations of Ecocrime. New York, Transnational Publishers Inc.

Westra, L., Bosselmann, K. & Westra, R. (2009). Reconciling Human Existence with Ecological Integrity. London, Earthscan.

Wetterstein, P. (1997). A Proprietary or Possessory Interest: A Condition Sine Qua Non for Claiming Damages for Environmental Impairment. In: P Wetterstein (Ed.), Harm to the Environment: The Right to Compensation and the Assessment of Damages (p. 29). Oxford, Clarendon Press.

WHAT (2000). Governance for a sustainable future, Report by the World Humanity Action Trust, London, World Humanity Action Trust.

Whelan, F.G. (1983). Democratic theory and the boundary problem. In: J.R. Pennock, & J.W.Chapman (Eds.), Liberal democracy (p.13). New York, New York University Press.

Wilkinson, D. (2002). Environment and Law. London, Routledge.

Williams, A .(2008). Turning the Tide: Recognising Climate Change Refugees in International Law. Law and Policy, 30, 502.

Williams, M. (2000). The uneasy alliance of group representation and deliberative democracy. In: W. Kymlicka, & W. Norman (Eds.), Citizenship in diverse societies (p. 124). Oxford, OUP Oxford.

Wind, M. (2009). The Commission White Paper. Bridging the gap between the governed and the governing?. Available at www.jeanmonnetprogram.org.

Wissenburg, M. (2004). Fragmented citizenship in a global environment. In: J. Barry, B. Baxter, & R. Dunphy (Eds.), Europe, Globalisation and Sustainable Development (p. 73). London, Routledge.

Wolf, J. (2007). The ecological citizen and climate change. In: Prepared for the workshop "Democracy on the day after tomorrow" at the ECPR Joint Sessions. Helsinki.

Wolfe, M. W. (2008). The shadows of Future Generations. Duke Law Journal, p. 1897.

Wonga, S., & Sharpb, L. (2009). Making power explicit in sustainable water innovation: re-linking subjectivity, institution and structure through environmental citizenship. Environmental Politics, 18, 37.

Wood, M. C. (2009). Advancing the Sovereign Trust of Government of Safeguard the Environment for Present and Future Generations (Part I): Ecological Realism and the Need for a Paradigm Shift. Environmental Law, 43.

Wood, P. (2000a). Intergenerational justice and curtailments on the discretionary powers of Governments. Environmental Ethics, 411.

Wood, P. (2000b). Biodiversity and democracy: rethinking society and nature. Vancouver, University of British Columbia Press.

Woodward, J. (1986). The non identity problem. Ethics, 804.

World Bank (1996). Assessing Aid: What Works. What Doesn't and Why. Oxford, Oxford University Press Inc.

World Commission On Environment and Development (1987). Our Common Future [Bruntland Report]. Oxford, Oxford Paperbacks.

Young, I.M. (1990). Justice and the politics of difference. Princeton, Princeton University Press.

Young, I.M. (1999). Justice, inclusion and deliberative democracy. In: S. Macedo (Ed.), Deliberative politics: essays on democracy and disagreement (p. 151). New York, OUP USA.

Young, I.M. (2000). Inclusion and democracy. Oxford, OUP Oxford.

Yourcenar, M. (1980). Les Yeux Ouverts. Paris, le Livre de Paris.

Zaccai, E. (2009). Développement durable: l'idéologie du XXIe siècle. Les Grands Dossiers des Sciences Humaines, n°14.

Zamagni, S. (1994). Global Environmental Change, Rationality and Ethics. In: L. Campiglio, L. Pineschi, D. Siniscalco, & T. Trevest (Eds.), The Environment After Rio (p. 235). Dordrecht, London, Graham & Trotman/martinus Nijhoff.

Zampetti, P.L. (1995). La democrazia partecipativa e il rinnovamento delle istituzioni. Genova, Edizioni Culturali Internazionali Genova.

Ziehm, C. (2005). Legal Standing for NGOs in Environmental Matters under the Aarhus Convention ad under Community and National law. Journal for European Environmental Planning Law, 287.

2 International sources

2.1 Treaties

Cartagena Protocol on Biosafety to the Convention on Biological Diversity, Jan. 29, 2000, 39 I.L.M. 1027;

Convention for the Protection of the Marine Environment of the North-East Atlantic, Sept. 22, 1992, 32 I.L.M. 1069

Convention on Access to Information, Public Participation in Decision- Making and Access to Justice in environmental Matters, Participants, June 25, 1998, 38 I.L.M. 517 (1999), *entered into force* Oct. 30, 2001.

Convention on Biological Diversity of 5 June 1992. 1760 UNTS, p. 79

Convention on Civil Liability for Damage Resulting from Activities Dangerous to the Environment, June 21, 1993, 32 I.L.M. 1228;

Convention on Cooperation and Sustainable Use of the Danube River, June 29, 1994, available at www.icpdr.org/ icpdr-pages/drpc.htm

Convention on Cooperation and Sustainable Use of the Danube River, June 29, 1994, available at www.icpdr.org/ icpdr-pages/drpc.htm

Convention on International Trade in Endangered Species of Wild Fauna and Flora, *3 March 1973* available at www.cites.org/

Convention on Long-range Transboundary Air Pollution, 13 November 1979. 18 ILM 1979, p. 1442

Convention on the Conservation of European Wildlife and Natural Habitats, 19 September 1979, ETS n. 104, UKTS n. 56 (1982)

Convention on the Protection and Use of Transboundary Watercourses and International Lakes, Mar. 17, 1992, 31 I.L.M. 1312

Convention on the Rights of the Child, 20 November 1989, 1577 UNTS 3

Convention on the Transboundary Effects of Industrial Accidents, Article 9, Mar. 17, 1982, 2105 U.N.T.S. 460

Convention to Combat Desertification in Those Countries Experiencing Serious Drought and/or Desertification, June 17, 1994, 33 I.L.M. 1328

Covenant on Civil and Political Rights, December 16, 1966, 999 UNTS, p. 171

Earth Charter, Mar. 2000, available at www.earthcharter.org/files/charter/charter.pdf

Energy Charter Treaty, Dec. 17, 1994; 33 I.L.M. 360

Framework Convention on Climate Change, May 9, 1992, 1771 UNTS 107, UN Doc. A/AC.237/18 (Part II)/Add.1

International Treaty on Plant Genetic Resources for Food and Agriculture, Nov. 3, 2001, available at ftp://ftp.fao.org/ag/cgrfa/it/ITPGRe.pdf

Kyoto Protocol to the United Nations Framework Convention on Climate Change, Dec. 11, 1997, 37 I.L.M. 22

Montreal Protocol on Substances that Deplete the Ozone Layer September 16, 1987 BGBl 1988 II, 1015; 26 ILM 1550 [1987]

North American Agreement on environmental Cooperation, Sept. 14, 1993, 32 I.L.M. 1480

Protocol Concerning Specially Protected Areas and Biological Diversity in the Mediterranean, June 10, 1995, 1999 O.J. (L 322) 3

Protocol on Water and Health to the 1992 Convention on the Protection and Use of Transboundary Watercourses and International Lakes, June 17, 1999, available at www.euro.who.int/ Document/Pehehp/ ProtocolWater.pdf

Protocol to the 1979 Convention on Long-Range Transboundary Air Pollution Concerning the Control of Emissions of Volatile Organic Compounds or Their Transboundary Fluxes, Nov. 18, 1991, 31 I.L.M 568

PRTR Protocol May 2003, in Kiev, Ukraine available at www.unece.org/env/pp/prtr.htm

Rotterdam Convention on the Prior Informed Consent Procedure for Certain Hazardous Chemicals and Pesticides in International Trade, Sept. 10, 1998, available at www.fco.gov.uk/Files/ kfile/CM%206119.pdf

Stockholm Convention on Persistent Organic Pollutants, Sept. 22, 2001, 40, I.L.M. 532

United Nations *Declaration on the responsibilities of the present generations towards future generations* UNESCO–United Nations Educational, Scientific and Cultural Organisation, New York 1997

World Charter for Nature, Oct. 28, 1982, 37 UN – GAOR, Supp. No. 51, p. 17, UN Doc. A/37/51

2.2 Other international sources

African Charter on Human and Peoples' Rights, adopted by the Organisation of African Unity, 27 June 1981, 21 I.L.M. 58 (1982)

Agenda 21 of the United Nations Division for Sustainable Development, adopted at the United Nations Conference on Environment and Development (UNCED), Rio de Janeiro, Brazil, 3-14 June 1992.

Bergen Ministerial Declaration on Sustainable Development in the ECE Region, 16 may 1990, UN Doc. A/CONF.151/PC/10; 1 Yearbook on International environmental Law 429 (1990): 4312

Communication ACCC/C/2008/32, submitted on 1 December 2008 by ClientEarth and others, *available at available at www.unece.org/env/pp/compliance/C2008-32/DatasheetC-2008-32v2009.01.19.doc.*

Conference on Security and Co-Operation in Europe Final Act of 1 August 1975, 14 ILM 1992, p. 1292.

Conference Report: "Resolution of the Avosetta Conference 11-12 October 2002, Amsterdam", *European environmental Law Review*, 2003, p. 34.

Declaration by the Environment Ministers of the region of the United Nations Economic Commission for Europe (UN/ECE), 4th Ministerial Conference „Environment for Europe", Aarhus, Denmark, 23-25 June 1998

Draft International Covenant on environmental and Development, 1995 and update in 2000 and 2004, Cambridge

ECOSOC, ECE, Meeting of the Signatories to the Convention on Access to Information, Public Participation in Decision-making and Access to Justice in environmental Matters, *Annex to the Addendum to the Report of the First Meeting of the Parties: Decision I/7 Review of Compliance* 4, U.N. Doc. ECE/MP.PP/2/Add.8 (Apr. 2, 2004), available at www.unece.org/env/pp/documents/mop1/ece. mp.pp.2.add.8.e.pdf

European Social Charter, adopted by the Council of Europe, 18 October 1961, entered into force 26 February 1965, European Treaty Series (ETS), No. 35

Fourth Ministerial Conference Environment for Europe, Århus, Denmark 23–25 June 1998 Declaration by the Environment Ministers of the region of the United Nations Economic Commission for Europe (UN/ECE), available at www.unece. org/env/efe/history%20of%20EfE/Aarhus.E.pdf

Johannesburg Declaration on Sustainable Development, adopted by the World Summit on Sustainable Development, 2-4 September 2002, U.N. Doc. A/ CONF.199/20, §13.

Joint Communique and Declaration on the Establishment of the Arctic Council, Sept. 19, 1996, 35 I.L.M. 1382

Ksentini Report, Final Report, UN Doc. E/CN.4/Sub.2/1994/9

Report of the second meeting of the Task Force on Access to Justice, UN Doc. MP.PP/WG.1/2004/3, 8 January 2004, *available at www.unece.org/env/ documents/2004/pp/mp.pp/wg.1/mp.pp. wg.1.2004.3.e.pdf,* Annex, p. 15, para. 17.

Report of the World Commission on Environment and Development: Our Common Future (Brundtland Report) of 1987, UN Doc. A/42/427, p. 40

Rio Declaration on Environment and Development, Report of the United Nations Conference on Environment and Development, Aug. 10, 1992, UN Doc. A/ CONF.151/26 (Vol. I)

Stockholm Declaration of the United Nations Conference on the Human Environment, UN – Doc. A/CONF. 48/14 (1972), reprinted in (1972) 11 ILM 1416

Turin Council: White Paper on the 1996 Intergovernmental Conference, 29 March 1996, Vol. II

U.N. Econ. & Soc. Council [ECOSOC], Econ. Commission' for Europe, *Report of the Meeting of the Parties to the Protocol on Water and Health to the Convention on the Protection and Use of Transboundary Watercourses and International Lakes*, U.N. Doc. ECE/MP.WH/2/Add.3, *available at www.unece.org/env/documents/2007/wat/wh/ece.mp.wh.2_add_3.e.pdf.*

3 Documents of the European Union

Directive 96/61/EC of 24 September 1996 Concerning Integrated Pollution Prevention and Control, OJ 1996, L 257/26

Directive 2001/18/EC of the European Parliament and of the Council of 12 March 2001 on the Deliberate Release into the Environment of Genetically Modified Organisms and Repealing Council Directive 90/220/EEC, OJ 2001, L 106/1

Directive 90/313/EEC of 23 July 1993 on the freedom of access to information on the environment provided the legal basis for access to environmental information in the EC countries and in other countries in the UN/ECE region, OJ L158/56

Directive 92/43/EEC of 21 May 1992 on the Conservation of Natural Habitats and of Wild Fauna and Flora, OJ 1992, L 206/7

European Commission's White Paper on European governance 2001

Detailed Table of Content

www.ingramcontent.com/pod-product-compliance
Lightning Source LLC
Chambersburg PA
CBHW080554030726
47589CB00003B/132